T0350167

Text Comparison and Digital Creativity

Scholarly Communication

Past, present and future of knowledge inscription

VOLUME 1

Text Comparison and Digital Creativity

The Production of Presence and Meaning in Digital Text Scholarship

Edited by

Wido van Peursen, Ernst D. Thoutenhoofd,
Adriaan van der Weel

BRILL

LEIDEN • BOSTON
2010

This book is printed on acid-free paper.

Library of Congress Cataloging-in-Publication Data

Text comparison and digital creativity : the production of presence and meaning in digital text scholarship / edited by Wido van Peursen, Ernst D. Thoutenhoofd, Adriaan van der Weel.
    p. cm. — (Scholarly communication, ISSN 1879–9027 ; v. 1)
    Contributions triggered by an international colloquium titled 'Text Comparison and Digital Creativity, an International Colloquium on the Co-production of Presence and Meaning in Digital Text Scholarship', held in Amsterdam on 30 and 31 October 2008 on the occasion of the 200th anniversary of the Royal Netherlands Academy of Arts and Sciences (KNAW).
    Includes index.
    ISBN 978-90-04-18865-5 (alk. paper)
    1. Criticism, Textual—Data processing. 2. Communication in learning and scholarship—Technological innovations. 3. Scholars—Effect of technological innovations on. 4. Electronic publications. 5. Manuscripts—Digitization. 6. Early printed books—Digitization. 7. Philology—Research—Methodology. 8. Bible—Criticism, Textual I. Peursen, W. Th. van II. Thoutenhoofd, Ernst D. III. Weel, Adriaan van der.

    P47.T43 2010
    801'.9590285—dc22

                           2010028034

ISSN 1879-9027
ISBN 978 90 04 18865 5

CONTENTS

List of Contributors vii

Foreword: Imagining the Manuscript and Printed Book in a
Digital Age ix
   Ray Siemens

Text Comparison and Digital Creativity: An Introduction 1
   Wido van Peursen

PART ONE

CONTINUATION AND INNOVATION IN
E-PHILOLOGY

In the Beginning, when Making Copies used to be an Art...
The Bible among Poets and Engineers 31
   Eep Talstra

Towards an Implementation of Jacob Lorhard's Ontology as a
Digital Resource for Historical and Conceptual Research in
Early Seventeenth-Century Thought 57
   Ulrik Sandborg-Petersen and Peter Øhrstrøm

PART TWO

SCHOLARLY AND SCIENTIFIC RESEARCH

Critical Editing and Critical Digitisation 79
   Mats Dahlström

The Possibility of Systematic Emendation 99
   John Lavagnino

The Remarkable Struggle of Textual Criticism and
Text-genealogy to become Truly Scientific 113
   Ben Salemans

PART THREE

CASE STUDIES

Seeing the Invisible: Computer Science for Codicology 129
  Roger Boyle and Hazem Hiary

Concrete Abstractions: Ancient Texts as Artifacts and the
  Future of Their Documentation and Distribution in the
  Digital Age 149
  Leta Hunt, Marilyn Lundberg and Bruce Zuckerman

Ancient Scribes and Modern Encodings: The Digital Codex
  Sinaiticus 173
  David Parker

Transmitting the New Testament Online 189
  Ulrich Schmid

Distributed Networks with/in Text Editing and Annotation 207
  Vika Zafrin

PART FOUR

WIDER PERSPECTIVES ON DEVELOPMENTS IN DIGITAL
TEXT SCHOLARSHIP

The Changing Nature of Text: A Linguistic Perspective 229
  David Crystal

New Mediums: New Perspectives on Knowledge Production ... 253
  Adriaan van der Weel

Presence beyond Digital Philology 269
  Ernst D. Thoutenhoofd

Indices
  Author Index 291
  Subject Index 294

LIST OF CONTRIBUTORS

R. (Roger) Boyle is Professor of Computing in the School of Computing at the University of Leeds, England.

D. (David) Crystal is Honorary Professor of Linguistics in the School of Linguistics and English Language at Bangor University, Wales.

M. (Mats) Dahlström is Assistant Professor in the Department of Library and Information Science, University College of Borås and Gothenburg University, Sweden.

H. (Hazem) Hiary is Assistant Professor in the Computer Science Department at the University of Jordan.

L. (Leta) Hunt is Associate Director of the inscriptiFact Project at the University of Southern California.

J. (John) Lavagnino is Reader in Digital Humanities in the Centre for Computing in the Humanities and the Department of English at King's College London, England.

M. (Marilyn) Lundberg is Adjunct Assistant Professor at Fuller Theological Seminar at Pasadena and Associate Director of the West Semitic Research and InscriptiFact Projects at the University of Southern California, US.

P. (Peter) Øhrstrøm is Professor of Information Science in the Department of Communication and Psychology at Aalborg University, Denmark.

D. (David) Parker is Edward Cadbury Professor of Theology and a co-Director of the Institute for Textual Scholarship and Electronic Editing at the University of Birmingham, England.

W.Th. (Wido) van Peursen is Associate Professor in Old Testament at the Leiden Institute for Religious Studies and Director of the Turgama project at Leiden University, the Netherlands.

B. (Ben) Salemans is Editor of www.neder-l.nl, the Netherlands.

U. (Ulrik) Sandborg-Petersen is Assistant Professor in Computational Linguistics at Aalborg University, Denmark.

U. (Ulrich) Schmid is Professor in the Evangelical-Theological Faculty and Researcher in the Institute for New Testament Research at the University of Münster, Germany.

R. (Ray) Siemens is Canada Research Chair in the Department of English at the University of Victoria, British Columbia, Canada.

E. (Eep) Talstra is Professor in Old Testament and Computer-Assisted Textual Research in the Faculty of Theology at the Vrije Universiteit of Amsterdam, the Netherlands.

E.D. (Ernst) Thoutenhoofd is Senior Researcher at the Virtual Knowledge Studio for the Humanities and Social Sciences of the Royal Netherlands Academy of Arts and Sciences, and University Lecturer in the Faculty of Behavioural and Social Sciences at the University of Groningen, the Netherlands.

A.H. (Adriaan) van der Weel is Bohn Professor in Modern Dutch Book History at Leiden University, the Netherlands

V. (Vika) Zafrin is Digital Collections and Computing Support Librarian in the School of Theology at Boston University, US.

B.E. (Bruce) Zuckerman is Professor of the Hebrew Bible in the School of Religion and Director of the West Semitic Research and InscriptiFact Projects at the University of Southern California, US.

FOREWORD:
IMAGINING THE MANUSCRIPT AND PRINTED BOOK IN A DIGITAL AGE

Ray Siemens

One of the most intriguing imaginings of the book and manuscript codex occurs when we endow it with human characteristics, when the book is something we perceive to have a body and a mind, a heart and a soul. Over time, we have both pleasurably noted and marvelled that the book at once *is embodied by us*—that is, we give it a physical shape, form, and content—and *is an embodiment of us*—acting, say, as our surrogate, across place and time. When we speak of the new and evolving electronic manifestation of the book today, however, we typically steer away from consideration of its more human characteristics. Instead, we have an awkward pattern of veering toward a characterisation of bookish manifestations that are quite foreign to us, focusing on the notion of an object that lacks recognizable physical form: disembodied texts, the ether of that which can be transmitted over computer networks, and so on. All told, this trend gestures more to the stuff of the grotesque and the futurist than that of the humanist, whose influence will surely be as strong in guiding the development of the book's current and emerging electronic manifestations as it has impacted, for centuries, the evolution of the manuscript and printed book. In our thinking about this, we need to chart a new course, one away from the futurist and the grotesque, and rather strongly toward that of the ongoing humanistic tradition of engagement with the codex form. Perhaps this should begin with a clearer understanding of our human relationship with the book itself.

The electronic representation of the codex—a book with no covers, as is said, perhaps with no immediately apparent physical form, lacking in assured fixity, and potentially with no beginning and no ending—can benefit from an understanding of the way in which aspects of the book have been figured variously across a number of centuries, centuries which we widely acknowledge to have seen two paradigm shifts: the first, from a culture that saw transmission of the codex via scribal copying to a culture that embraced the mechanical processes

of the printing press, and the second, which is currently well along
its way in seeing transmission shifting from the mechanical processes
of the printing press to that of computer-assisted forms enabled by
electronic media and the internet. My entry will be via several vignette-
like imaginings of the codex form alongside its human producers and
consumers, imaginings that imbue the book and manuscript with a
reflected humanity and personality that is, realistically, denied by its
physical limitations but are suggestive of the relationship between
physical form and intellectual content, between presence and meaning,
in both the physical book form and its electronic counterparts.

A context for theorising the codex

Our attention to the physical elements of books and reading is omni-
present, though often taken for granted. Indeed, many of us only
consciously engage the physical form of the book when we encounter a form that somehow challenges our expectations; to list a few
examples: the young child's first encounter with the book, which most
likely involves the sense of taste at least as much as sight; the paper-
back that has permanently-bent pages and, thus it sits awkwardly; the
book that is, say, circular in shape (rather than rectangular), or with
a fold-out cover; and the book which has a page layout or font that is
unique, such that our reading may be thrown off-balance for the first
few minutes. Such encounters we have with books provide ample evidence of our engagement with the codex as a physical object as much
as one housing intellectual content, and this separation of the codex's
physical and intellectual attributes is something that has seen remark,
exploration, and even good exploitation for some time.

Inscription and the human codex

Beyond the anecdotal, one finds this exemplified well in early manuscript culture: where the connection of the book with the physicality
of its production, inscription, and use is much closer than it could
possibly be in an age of comparatively hands-off mechanical reproduction—and even closer than it might be in early print culture, which
adopted aspects of the scribal codex form at the same time as it sought
new formal expressions, ones at times made more convenient or possible by the technologies of mechanical reproduction. Specific to my
purpose here is the notion of inscription as it relates to that time,

to the manuscript, and to the body. When we say, today, that we've inscribed something, we most likely mean that we've put pen or pencil to paper and written, yet an earlier interpretation of inscription would have the potential to be much more allusive.[1] Beyond the marks of the scribe on the page, one might equally consider the inscription left by one's feet on the ground, or, by way of further example, the marks of violence inflicted on a martyr (see Plate 1). The footprint and the wound are each a type of inscription that have clear association with the marks on a scribe's page. Each of the inscribed media, if you will— the earth, the skin of Christ, and the prepared skin of the animal out of which most medieval manuscripts are made—bears the mark of the inscriber, in an act associated as much with the consequence of making that mark (walking, infliction of life-threatening wounds, or writing) as with what the mark suggests and represents (perhaps flight, martyrdom, or the word of God).

The codex was a bodily entity for the early reader, one might say, and the acts of reading and writing were as much dominated by intellectual activity as by conscious sensual experience, even at times depicted as sexual. For those who produce it and for those who consume it, the book is an object very much associated with the humanity and human desire that is analogically imbued via its creation and use. The book's physical form and sensuality is seen to resemble ours, the content with which we imbue it is wholly ours, and the actions associated with its creation and use are human as well.

Content and meaning inscribing book form: The book of the heart

Ideas associated with the notion of the *book of the heart* illuminate this further, one of the best-known exemplifications of this seen in a fresco from the Cathedral of Albi (c. 1490, see Plate 2), depicting the last judgement with human figures in which their hearts are represented as opening like books.[2] Founded on the idea that one's innermost thoughts and personal history could be considered as inscribed within us, this tradition sees depiction across several millennia as a

---

[1] Following, here and below, Warwick Frese and O'Brien O'Keefe, *The book and the body*, Carruthers, 'Reading with attitude', Camille, 'The book as flesh and fetish', and others.

[2] Following Jager, *Reading the book of the heart*, and others, here and below.

scroll rolled up in one's heart, a wax tablet which could be wiped clean by conversion to Christianity and, in the codex form, as a metaphoric reference to one's life record, connoting also intellectual capacity to recall, comprehend, and enact one's moral sense. Consider, for example, the personal associations of the heart-shaped book in the much-copied work by the Master of the View of Saint Gudule (see Plate 3). While the physical form of the heart materially reflects an essential intellectual frame of reference, it is not the physicality of the book that is of paramount importance here; rather, form here readily depicts for us content, such that we can begin to understand content via observation of form. Our body here is represented by its most significant element privileged by this tradition.

Formal presence, illuminating meaning and practice: Herbert's 'Easter Wings'

While textual theory of the past several decades has taught us not to ignore the importance of form, most will concur that form is rarely an end in itself and that medium is not exclusively the message, as much as medium may have everything to do with its conveyance. Rather, form and content, meaning and presence, are correlative, each informing the other. The seventeenth-century English devotional poet George Herbert illustrates such a relationship in his poem 'Easter Wings', which has captured the professional imagination of those involved in textual studies, and editing, for over a century. The poem resists a traditional pattern of reading in its early printed form, in the first instance because one has to change the orientation of the text in order to begin to engage its textual content. Plate 4 (showing the second edition) depicts the text as it appears when one holds the book in the traditional manner. It must be re-oriented for the text to be read in normal fashion. Should one choose not to turn the book, one treats the content graphically rather than textually and appreciates the image of a pair of wings suggested by the correctly-oriented title. If one does choose to turn the book after recognising the shape, one can engage it textually and, in doing so, might note that the poem is connected as the title suggests to the church calendar and treats, among other things, the ascension of Christ and all that it symbolises.

In changing the orientation of the book, one also realises via its content that 'Easter Wings' is a poem the form of which plays a very important role beyond that of the wing shape resembled by the text. Each ten-line stanza represents a decline and an elevation, with the decline emphasised in the shortening of the poetic lines, which move towards the phrases 'most thin' and 'most poor' in the middle of each stanza. The rise is indicated by increasing line lengths from the centre of each stanza to its end. When those expert in Herbert's thought discuss 'Easter Wings', most work towards discussion of the notion expressed in the second-last line of page 35—'if I imp my wing on thine'—and treat it as a plea of the speaker, who presents the poem as a prayer for his own rise with Christ's ascension.

This pattern of interpretation is significant, to be sure, but its significance is increased by a clearer understanding of the full richness of interaction between presence and meaning in this text. It is generally accepted that each of the stanzas have association by their first word with, respectively, the speaker (on page 35 of the original; the first to be read in Plate 4) and (on page 34; the second to be read) the Lord, and it is considered that the poem is a prayer that can be intended for actual use, as are others in Herbert's collection of *The Temple* that houses them. Further, it has been suggested that devotional practices in Herbert's time would see a set prayer (such as this one, perhaps) read, silently alone or aloud to a group, followed by a time of silent deliberation in which the prayer book would likely be closed. Noting these points when exploring the relationship of form and meaning makes available to us some pertinent considerations. The first is that the reading process encourages holding the book in a specific manner, making it necessary to turn the volume in such a way that the poem can be read, and requiring both hands to be on the volume; one on each page and cover. The second is that, after the poem is read, the book would likely be returned to its more usual alignment and then, with two hands still on the pages and covers, the volume would be closed. The third takes place at the closure of the text—an act that many say brings the greatest formal significance to the content of the piece, because in closing the book in the manner that many believe it would have been closed, two very important events take place: one is that the two wings on opposite pages—that associated chiefly with the speaker and that with the Lord—are 'imped', are brought together, in the way called for by the prayer; the other is that the most natural position of the closed hands on the volume as the book is closed is that resembling the position of prayer. Ingeniously, form encourages action and reading practice in ways central to that which is treated by content.

Multiplicity of intellectual presence and its formal representation:
A comment on Peter Lombard's Commentary

Without engagement of the poem's form, in the case of 'Easter Wings' the reader cannot even begin to engage the textual content; but, once engaged, it is clear that form and content serve each other cleverly. Academic tradition might hold that the more salient of the two—that associated with form and physical presence and that associated with content and meaning—is intellectual content, but this example and others suggest the importance of remaining cognisant of the salient strength of form and presence. When the book, which can be figured as an extension of ourselves, speaks for us and encourages us to act, formal concerns manifest at least as importantly as that which form can hold and communicate for us. In this regard, one last print-oriented artefact to consider is presented in Plate 5, a page originating in the scriptorium of Peter Lombard, Bishop of Paris, exemplifying a contribution to the tradition of representing the complexly-interwoven commentary associated with the Epistles.

Here, in a method which serves as a recognisable foundation for that employed in scholarly editions today, extensive commentary is arranged to surround one small section of one epistle, the formal considerations of the page mimicking and augmenting, visually, aspects of the intellectual engagement by those whose expertise is rendered in the commentary on the section residing at the page's centre. The composite authorial form is provided by a line that runs almost the full vertical length of the inscribed page's right side, from head to feet of the human representation, at times marked with indications relating original text to the originators of the commentary (Jerome, Augustine, Ambrose) that provides the intellectual context for the reader's consideration. The visual image both identifies those providing the commentary and suggests clearly the composite, corporal nature of commentary in relation to the central author's body of work; the visualised body represents the composite sum of its constituent, intellectual parts—much as the way in which the material on the page is comprised of the original text itself plus that which tradition has deemed essential toward the understanding of this original text.

Presence and meaning, in the absence of physical form

Turning from argumentative preface to the matter of the collection, it is worth noting that the leap forward towards e-science and e-humanities is nevertheless not so great that one can no longer understand the future of disciplinary endeavour as anything other than a forward progression built on past advances, extending extant intellectual tradition. For all the radical rhetoric which surrounds the move from print to electronic media, even this paradigm shift isn't so radical that history and tradition are fully erased or rendered wholly irrelevant. Indeed, a valuable perspective for understanding our current paradigm shift is via the lens provided by that of the movement from manuscript culture to print culture, from scribal reproduction to mechanical. Our adoption and understanding of electronic media in those book-related activities that we carry out professionally and personally is best informed, explicitly and analogically, by our understanding of a past that presents valuable perspectives on and prospective solutions to concerns arising as part of our engagement of electronic media, specifically with reference to the patterns of physical and cognitive representation and the transmission, as well as the interaction, that they facilitate.

One does not have to look beyond the contents of this volume, and its resonance into all corners of the fields drawn upon and reflected among its papers, to see a very clear recognition that humanities-derived textual technologies and methodologies drawn from the age of manuscript and print have ready application to our understanding of the evolving new media culture in which we live, in the e-humanities, and even in emerging e-sciences. Examples abound, in the pages of this collection and well beyond, the most striking along these lines perhaps being the adoption of techniques associated with complex textual collation by those doing string-based analysis of DNA; equally striking is the connection between twelfth-century attempts at rendering detailed, human-associated commentary with that of our own evolving implementations of commentary in social computing environments that very clearly connect our own contributions to argument and discussion with visual self-representations, and the way in which we are beginning to imagine our human relationship to the form and style of the physical computer as well as its on-screen interface—particularly in increasingly ubiquitous devices such as laptop computers, iPhones, and emerging micro-computers and reading tablets.

As with the book over time, so, too, with these devices. The devices that we embody by giving shape, form, and content continue to act as embodiments of us. Study of the multiple-authored disembodied text as represented by Lombard offers an important starting point in such considerations, as does our understanding of Herbert's method in manipulating physical form to augment significantly the reader's devotional experience, as does engaging the book of the heart tradition and the nature of the form–content relationship there—and, I would urge overall, as does our consideration of the relationship we have had, over time, with the book's physical form. Indeed, whatever the book might be—in manuscript, print, or some electronic manifestation—its connection is at least as much with us as humans as it is with the technologies associated with its production and dissemination, and surely more. Its sensuality is seen to resemble ours, the content with which we imbue it is ours, and the originating actions associated with the creation and use of the book are ours as well.

Nevertheless, even working in a tradition that understands the relationship between content and form, meaning and presence, we are often surprised when the results of our technological translation from print to digital representation yields unexpected results—largely, I would argue, based on a seemingly unconscious extension of our print expectations to the electronic medium and, further, by our tradition's valuation of intellectual content over physical form. A focus in our attention, now, on what is lost in that media-technological translation, on what from manuscript and print culture cannot yet be readily rendered and modelled in electronic media, provides us with a crucial foundation for research endeavour that gets to the heart of the paradigm shift and encourages an awareness of the way in which the electronic medium reconfigures our relationship to text at all levels. Examination of this crux requires exactly what is championed by the work represented in this volume: a firm grounding in the traditions associated with textual history and its culture coupled with the intellectual dexterity and interdisciplinary foresight to imagine how such a history can be extended into the world of electronic media. Chief among pertinent concerns here are issues of representation and re-presentation—and a focus on the maintenance of the conjoint importance of form and content, of presence and meaning, and the creative patterns that we have brought and continue to bring to our design and our intellectual engagement of the textual artefacts that show no sign of wavering in their centrality.

References

Camille, Michael, 'The book as flesh and fetish in Richard de Bury's *Philobiblon*', in Warwick Frese and O'Brien O'Keefe (eds), 34–77.

Carruthers, Mary, 'Reading with attitude, remembering the book', in Warwick Frese and O'Brien O'Keefe (eds), 1–33.

Jager, Eric, *Reading the book of the heart from the Middle Ages to the twenty-first century* (Chicago, 2002).

Muri, Alison, 'The electronic page. Body/spirit, the virtual-digital and the real-tactile', in *Architectures, ideologies, and materials of the page* (Saskatoon, 2002).

Warwick Frese, Dolores, and Katherine O'Brien O'Keefe (eds), *The book and the body* (Notre Dame, 1997).

——, 'Introduction', in Warwick Frese and O'Brien O'Keefe (eds), ix–xviii.

TEXT COMPARISON AND DIGITAL CREATIVITY: AN INTRODUCTION

Wido van Peursen

What are the effects of digital transformations in text culture on textual scholarship? What rules and guidelines are appropriate for the digital interpretation of text? What 'virtual' values do we turn to as the object of digital humanities scholarship? What is the role of viewpoint, language, tradition and creativity in quantitative text comparison? What connections exist between textual scholarship, interpretation, and e-infrastructures for research?

These questions were addressed at an international colloquium held in Amsterdam on 30 and 31 October 2008 on the occasion of the 200th anniversary of the Royal Netherlands Academy of Arts and Sciences (KNAW).[1] The conveners, Ernst Thoutenhoofd and the undersigned, represented two institutions: the Virtual Knowledge Studio for the Humanities and Social Sciences (VKS), a programme of the KNAW; and Turgama, a research project at the Institute for Religious Studies of Leiden University, funded by the Netherlands Organisation for Scientific Research (NWO). The VKS aims to support researchers in the creation of new scholarly practices. Turgama deals with computational linguistic and comparative analysis of ancient texts. The colloquium was given the title 'Text Comparison and Digital Creativity. An International Colloquium on the Co-production of Presence and Meaning in Digital Text Scholarship'. The contributions to the present volume were triggered by this KNAW colloquium.

The present chapter gives an introduction to the theme of the volume and explains what we mean by 'text comparison', 'digital creativity', 'presence' and 'meaning', and why these words figure in the title of this volume. It also highlights some of the issues that emerged in the colloquium discussions as being crucial for a proper assessment of the transformations in digital text scholarship, such as the question as to

---

[1] We also gratefully acknowledge the support of the KNAW for the colour section of the present volume.

the status of digital research objects used in e-philology. This introductory chapter is followed by four sections, each of which explores the theme from a different perspective.

Text comparison and digital creativity

Within the broad field of 'digital humanities', the present volume zooms in on philology, which can be defined as 'the study of literature, in a wider sense, including grammar, literary criticism and interpretation, the relation of literature and written records to history, etc.' (*OED*). Philology provides a fascinating case study of the transition from 'traditional' research to computer-based research in humanities scholarship, because of the textual nature both of its objects of research and of the vehicle of scholarly communication and knowledge representation (see the section 'knowledge creation and representation' below). The present volume contains contributions from various subdisciplines of philology, but the chief focus is on biblical studies. This branch of philology has a longstanding history of traditional scholarship, witnessed some of the earliest earliest attempts to create annotated databases, and is still an area in which pioneering initiatives in e-philology are taking place.

Text comparison, which figures in the first part of the title of the present volume, is at the very heart of philology. In textual scholarship we need instruments that allow us to compare various literary texts. The comparison may involve various sections of the same text (cf. Talstra's reference to concordances),[2] texts that are in some way 'parallel' (cf. Talstra's reference to synopses), and texts that are more loosely related, containing, for example, motifs and themes that frequently recur in certain cultures, or other intertextual relationships. Parallel texts, texts that in one way or another are witnesses to the same composition, can be described as a continuum that 'ranges from the extreme of extensive verbatim quotation, on the one hand, to the point where no relationship is discernible, on the other'.[3]

The philological activity of text comparison goes back to Antiquity. One of the means to compare texts was to present them in parallel columns, a usage that is attested, for example, in Origen's third-century

---

[2] An author's name between brackets refers to the author's contribution to the present volume.

[3] Wise, *A critical study of the Temple Scroll from Qumran Cave 11*, 207; cf. Van Peursen and Talstra, 'Computer-assisted analysis of parallel texts in the Bible'.

Hexapla, which presented six versions of the Old Testament (including the Hebrew text, a transliteration of the Hebrew in Greek characters, and four Greek translations) in parallel alignment. After the invention printing, various Polyglots were created, such as the Complutensian Polyglot (1514–1517), the Antwerp Polyglot (1569–1572), the Paris Polyglot (1629–1645) and the London Polyglot (1654–1657), which contained the text of the Bible in different languages (Hebrew, Greek, Aramaic, Syriac, Latin, Arabic). These multilingual editions gave an important impetus to the text-critical and philological study of the Bible. For the poster of the colloquium that triggered the contributions to this volume, we used a picture of a beautiful Polyglot Psalter from the Library of Leiden University (see Plate 6 and cover). This twelfth-century manuscript gives the Psalms in Hebrew, Latin (the version 'iuxta Hebreaos'), Greek and again Latin (the 'Psalterium Gallicanum').

The focus of the present volume, however, is neither on text comparison in Antiquity, nor on the production of Polyglots in the first centuries after the invention of printing, but rather on text comparison and philology in the digital age. This brings us to the other part of the title of this volume, 'digital creativity'. In its original use as the opposite of 'analogue', 'digital' refers to the mathematical representation of information and involves the conversion of continuous information into discontinous symbols, such as alphanumeric characters. Although various digital systems have been developed in the course of history, nowadays 'digital' is mainly associated with the binary digital system used in computer technology. People frequently use, for example, 'digital humanities' and 'humanities computing' as mere synonyms. At the same time, however, 'digital' seems to surpass its assocation with calculation and computing, because in modern digital tools the binary digital system is often hidden in a black box and its users may be hardly aware of it. People who use the Internet for social networking, for example, will not readily associate this networking with calculation and binary information systems. As such, 'digital' has become a designation of electronic information and communication technology and the processes and techniques related to it.

'Digital' relates to 'creativity' in various ways. First, when we use 'digital' to describe the new ways in which information can be stored and represented, we can say that the digital media provide new possibilities for knowledge representation that go far beyond the potential of analogue data. It requires creativity to discover these new opportunities and to go beyond just using the digital tools and media in the same manner as we use the analogue instruments. Secondly, if 'digital' is used to refer to electronic communication, its relation to creativity is that digital instruments offer new chances for networking and cooperation, and as such support new creative processes. Thirdly, when 'digital' is used to indicate the mathematical representation of information and the use of the calculation power of the computer to handle and investigate it, the relation between 'digital' and 'creativity', becomes a methodological question: how do computation, calculation and sorting of data relate to traditional values of scholarship that seem to escape calculation, such as subjectivity, the master's eye, and the intuition of the experienced scholar?

Humanities scholars who are sceptial about the potential of the computer as a research instrument for philology may see the relationship between 'digital' and 'creativity' mainly as the opposition between two modes of scholarship, quantitative approaches which are useful in the sciences and qualitative research that is typical of humanities scholarship. As we shall see, however, the interaction between human researchers and computers, and between calculation and interpretation, is far more complex. Applications of IT in, for example, bioinformatics, in which the computer interacts with the human researcher to extract knowledge from immense data collections, or studies in the field of artificial creativity, show that the 'computational' and the 'creative' do not exclude each other. Definitions of creativity often include the notions of plurality and variation. Working with huge amounts of data—even, or perhaps even more, when they are handled in interaction with IT—requires creativity. The real challenge is 'to capture the messy complexity of the natural world and express it algorithmically' (Boyle and Hiary, quoting Teresa Nakra, who calls this 'the Holy Grail of computer science').

The title of this volume brings 'text comparison' and 'digital creativity' together. Its aim is to explore recent developments in comparative textual scholarship brought about by the use of IT.

Presence and meaning

Presence, meaning, and interpretation

'Presence' and 'meaning' figure in the subtitle of this volume. Traditionally textual scholarship is concerned with interpretation, the attribution of meaning. In *The production of presence. What meaning cannot convey*, Hans Ulrich Gumbrecht uses the terms 'presence' and 'meaning' to refer to the role of materiality ('physics') and meaning ('meta-physics') in textual scholarship. 'Presence' refers to the materiality of things, to the physical aspects, to 'what meaning cannot convey'. 'Meaning' refers to that which goes 'beyond' what is physical. It is related to metaphysics, which Gumbrecht describes as 'an attitude that gives a higher value to the meaning of phenomena than to their material'.

Gumbrecht does not reject interpretation (unlike some other authorities, see below), but his main concern is the rehabilitation of 'presence'. He argues that Descartes' dichotomisation between 'spiritual' and 'material' led to the introduction of the subject/object paradigm, in which 'to interpret the world means to go beyond its material surface or to penetrate that surface in order to identify a meaning (i.e., something spiritual) that is supposed to lie behind or beneath it'.[4] He describes this development as follows:

> Medieval Christian culture was centered on the collective belief in the possibility of God's real presence among humans and in several rituals, most prominently the Mass, that were meant to constantly produce and renew such real presence. (...) In modern culture, in contrast, beginning with the Renaissance, representation prevails over the desire for real presence. Representation is not the act that makes 'present again', but: those cultural practices and techniques that replace through an often complex signifier (and make thus available) as 'reference' what is not present in space or time. (...) My innovative thesis lies in the claim that, ever since the historical moment that we call the 'crisis of representation', around 1800, our culture has developed a renewed longing for real presence.[5]

Although Gumbrecht sketches a long pre-history of the re-awakened interest in 'presence', we think that it is only since the second half of the twentieth century that the hegemony of meaning has been threatened seriously. Interpretation, the academic, analytical and intellectual activity that concerns the attribution of meaning, has come under attack. In her 1966 essay 'Against interpretation'[6] Susan Sontag calls interpretation 'the intellect's revenge upon art'. The final words of her essay are: 'in place of a hermeneutics we need an erotics of art'. Robert Alter speaks of 'the heresy of explanation'[7] and claims that 'the

---

[4] Gumbrecht, *Production of presence*, 26.
[5] Gumbrecht, *Powers of philology*, 11.
[6] Cf. Gumbrecht, *Production of presence*, 10.
[7] Alter, *The five books of Moses*, xvi; cf. Talstra's remark on the 'interpretative extras' in Bible translations.

unacknowledged heresy underlying most modern English versions of
the Bible is the use of translation as a vehicle for explaining the Bible
instead of representing it in another language',[8] the result being that
'the modern English versions—especially in their treatment of Hebrew
narrative prose—have placed readers at a grotesque distance from the
distinctive literary experience of the Bible in its original language'.[9]
George Steiner speaks of the 'Byzantine dominion of secondary and
parasitic discourse over immediacy, of the critical over the creative'.[10]
And Gumbrecht 'challenges the broadly institutionalised tradition
according to which interpretation is the core practice, the exclusive
core practice indeed, of the humanities'.[11]

The 're-awakened interest in presence' is visible in the attention
paid to texts as artefacts and the material aspects of the carriers of
texts, which hardly receive any attention in traditional textual scholarship. In the field of biblical scholarship, for example, most standard
works on textual criticism hardly pay any attention to mise-en-page,
delimitation markers, paleography, and codicology. In the last decades
this situation has changed and various research groups dealing with
these aspects of textual transmission have been established, such as
the Pericope Research Group (focusing on unit delimitation in biblical manuscripts)[12] or the COMSt Network (dealing with the interdisciplinary study of Oriental manuscripts).[13] Besides the understanding
of a text as *text*, that is, 'as reduced to linguistic sign sequences', there
is the concept of text as *document*, that is, 'as a meaning conveyed by
those linguistic sequences in conjunction with layout, typeface, colour
and the rest of the graphical and material appearance that the document provides' (Dahlström; cf. also Talstra on the distinction between
'text' and 'document').

Dissatisfaction with the analytical academic attribution of meaning
has led to suggestions to avoid distance-creating interpretation and to
strive for immediacy. However, opinions differ about what this immediacy is. For Gumbrecht it is the material, physical presence of the
objects of the world. This reminds us of Walter Benjamin, who speaks
of the perception of an object's aura, that is: its unique existence in

---

[8] Alter, *The five books of Moses*, xix.
[9] Alter, *The five books of Moses*, xvii.
[10] Steiner, *Real presences*, 38.
[11] Gumbrecht, *Production of presence*, 1.
[12] See http://www.pericope.net.
[13] http://www1.uni-hamburg.de/COMST.

time and space. For Walter, too, presence ('Das Hier und Jetzt des Originals') is a prerequisite of authenticity ('Echtheit').[14] In his essay 'Das Kunstwerk im Zeitalter seiner technischen Reproduzierbarkeit' ('The work of art in the age of mechanical reproduction'), he argues that the distance or unapproachability ('Unnahbarkeit') of a work of art cannot be overcome by mechanical reproduction because an object's aura cannot be reproduced.[15] For Steiner, however, the mystery of 'real presence' means having access to that which transcends the physical reality. For Alter the immediacy is achieved by the appreciation of the artistic value of the literary work, including all its ambiguities that philologists try to explain away in their quest for clarity. Some readers of the Bible justify their avoidance of the analytical interpretation and their search for direct, spiritual access to the biblical text through the centuries-old tradition of the *lectio divina*.[16]

Both Gumbrecht and Steiner advocate 'Real Presences'. Both will agree that going to a concert where Händel's *Messiah* is performed gives a better access to that work than any musicological analysis of it and that visiting the National Gallery of Berlin to see Caspar David Friedrich's *Monk by the Sea* gives a more direct access to this painting than reading an interpretation of it by an art historian. And since also every mechanical or digital reproduction is necessarily always only an interpretation (see below, 'The creation of digital research objects') we could add with Benjamin that a mechanical reproduction of a performance of the *Messiah* or a digital image of the *Monk by the Sea* also lacks presence.

When, however, it comes to texts, the differences between the views of Gumbrecht and Steiner come to the front. For Gumbrecht, 'presence' refers to the spatial relationship to the world and its objects and hence to the material carriers of texts. In Steiner's view the effect of the confrontation with the primary (for example a literary text) is twofold: it gives direct access both to the physical material, avoiding the detour of interpretation, and to that which goes beyond description and interpretation, namely the transcendental real presence.

---

[14] Benjamin, 'Das Kunstwerk', 352: 'Das Hier und Jetzt des Originals macht den Begriff seiner Echtheit aus'; English translation (Benjamin, *Illuminations*): 'The presence of the original is the prerequisite to the concept of authenticity' (220).
[15] See especially the *Zweite Fassung* in *Gesammelte Werken* 1/2 [see below, note 20], 480, note 7.
[16] Cf. Reedijk, Zuiver lezen.

Presence, meaning, and digital textual scholarship

We included 'presence' and 'meaning' in the title of this volume, because one of the things that requires further investigation is the question as to how 'presence' and 'meaning' relate to digital textual scholarship. On the one hand, digital textual scholarship has to face the challenges to interpretation put forward by Steiner and others. According to Steiner, 'what looks to be certain is that the criteria and practices of quantification, of symbolic coding and formalisation which are the life breath of the theoretical do not, cannot pertain to the interpretation and assessment of either literature or the arts.'[17] This claim provides a challenge to the analytical and mathematical approaches that characterise the computational analysis of texts, even more than to traditional philology. It raises the question as to how the computational, analytical work done in digital scholarship relates to the subjective moods of interpretation and intuition that characterise traditional philology.[18]

On the other hand, if our claim is correct that 'the renewed longing for real presence' increased considerably in the second half of the last century, this development overlaps with the technological innovations of the digital age, which made new realisations of 'presence' possible. The new technologies support the study of texts as artefacts, enable the 'presentification'[19] of research objects on the computer screen, and help discover what stories the messenger can tell beyond the message it carries (Boyle and Hiary). The present volume provides a number of examples of this role of the new technologies.[20] The digitisation of textual research objects, in turn, requires reflection on the relation-

---

[17] Steiner, *Real presences*, 79. But Steiner does not completely reject the philological analysis. He refutes F.R. Leavis' claim that 'linguistics has nothing to contribute to the understanding of literature', because, 'an informed alertness to the phonetic, lexical, grammatical instrumentalities of a text both disciplines and enriches the quality of interpretative and critical response' (84). 'Therefore radical doubts, such as those of deconstruction and of the aesthetic of misreadings, are justified when they deny the possibility of a systematic, exhaustive hermeneutic, when they deny any arrival of interpretation at a stable, demonstrable singleness of meaning. But between this illusory absolute, this finality which would, in fact, negate the vital essence of freedom, and the gratuitous play, itself despotic by its very arbitrariness, of interpretative nonsense, lies the rich, legitimate ground of the philological' (164–165).

[18] See the section 'Scholarly and scientific research' below.

[19] Gumbrecht, *Production of presence*, 94.

[20] See the contributions by Boyle and Hiary, Parker, Hunt, Lundberg and Zuckerman, and Schmid.

ship between texts and their carriers. An artefact may be the carrier of various texts or various artefacts may together contain one text (Hunt, Lundberg, Zuckerman; cf. above, on the distinction between 'text' and 'document' made by Dahlström and Talstra).

Presence of digital research objects

Having observed that new technologies enable the 'presentification' of research objects, we have to address the question as to what sort of presence is created by digital technologies. Do high-resolution images of an ancient inscription indeed realise its 'presence'? The claim that they do and that they provide immediate access to the objects they (re-)present,[21] implies the identification of the real object, say, in a museum or in an archeological site with its digital representation in, say, an online repository. This identification can be challenged. Does the digital representation of a research object have the same 'presence', 'authenticity' or 'aura' as the object it represents? Or should we, with Benjamin, reserve 'presence' for an object's unique existence in a certain time and place, which cannot be reproduced?[22] In Benjamin's view, the desire to bring things 'closer', has led to the acceptance of reproduction,[23] but 'even the most perfect reproduction of a work of art is lacking in one element: its presence in time and space, its unique existence at the place where it happens to be'.[24] Benjamin's observations in the first half of the twentieth century on 'mechanical reproduction' apply very well to the 'digital reproduction', at the beginning of the twenty-first century.

---

[21] Cf. above, the section 'Presence, meaning, and interpretation' on Gumbrecht's distinction between 'presentation' and 'representation'.

[22] Benjamin, 'Das Kunstwerk', 352.

[23] Benjamin, 'Das Kunstwerk', 355: 'Die Dinge *näher* zu bringen, ist ein genauso leidenschaftliches Anliegen der gegenwärtigen Massen, wie es ihre Tendenz einer Überwindung des Einmaligen jeder Gegebenheit durch die Aufnahme von deren Reproduktion darstellt.' [The *Zweite Fassung* of this essay in idem, *Gesammelte Schriften* 1/2 (p. 479) reads: 'Die Dinge sich *räumlich und menschlich* "näherzubringen"...' (italics mine); on the various versions of Benjamin's essay see the editors' remark in Volume 7/2, 661.]

[24] Benjamin, 'Das Kunstwerk', 352: 'Noch bei der höchstvollendeten Reproduktion fällt *eines* aus: das Hier und Jetzt des Kunstwerks—sein einmaliges Dasein an dem Orte, an dem es sich befindet'.

The relationship between the digital objects and the real objects they allegedly represent is not only very complex,[25] it also develops over time. In this development the digital objects seem to evolve from mere secondary representations of traditional research objects to new kinds of research objects in themselves. In scholarly practice the 'original' and 'real' objects run the risk of being delivered to oblivion, because digital representations have taken their place.

A telling example of this development may be the way in which the Codex Sinaiticus was, is, and will be consulted. As Parker notes, access to the primary materials has always been restricted and visiting the four locations over which they are scattered is 'a formidable logistical undertaking'. The Digital Codex Sinaiticus provides a solution to these problems: access is no longer restricted and logistic problems have overcome. But now that the digital object has become available, who will ever go back to the 'real' manuscript? And if one does so, and arrives at new readings or new insights, will the claim of one single scholar who has consulted the original manuscript outweigh the judgment of dozens of colleagues who base their observations on the high-quality digital representation of it? The real danger is that this single scholar will not be taken seriously because the community at large has not the resources to perform autopsy.

The same question applies to electronic collections of ancient corpora, in, for example, the Perseus project, the Comprehensive Aramaic Lexicon, or the Stuttgart Electronic Study Bible. Although the digital objects are most often still conceived as presentations or representations of 'real' data (objects in museums, printed text editions), more and more often they receive a status of their own and become independent of the realities they represent. Will the next generation of scholars still take the trouble to check printed critical editions (which, admittedly, are representations as well) for Greek and Latin texts that can be found on the internet? Or will, for example, the Perseus collection receive a status as the place where *the* classical literature can be found? Or will the digital (re)presentations be regularly updated? If they are inevitably interpretative (cf. Dahlström), then like translations

---

[25] Cf. Hunt, Lundberg and Zuckerman in this volume: 'In order to provide a solid foundation for intuitive access, one must establish and facilitate a strong relationship between the model and the "real world". The problem is that it is sometimes not altogether clear what the "real world" is, and this "reality-concept" shifts remarkably depending on one's perceptions.'

or text editions they will need to be reperformed every twenty-five years or so—or perhaps even more frequently, given the tremendous speed at which digital technologies develop.

Similar questions can be raised regarding secondary literature. Google Book Search opens up new horizons for the accessibility of scholarly literature. But what are the effects? Will digitised sources that can easily be accessed by a simple search command be quoted more frequently than sources that exist only in printed editions and for which someone needs to go to the shelves of the library? And will people be sufficiently aware that the quality and accuracy of Google Book Search leaves much to be desired (Dahlström)?

An interesting analogy to the tendency that the digital research objects become increasingly independent from the 'real' objects they represent is provided by the digital representations of the 'real' world in virtual words such as Second Life and the way in which virtual objects function and develop there.[26] In Second Life we can observe how virtual objects that started as imitations of 'real' objects increasingly acquire independence in relation to the objects that they originally imitated. It is to be expected that also digital research objects of philology, such as those in the database of the West Semitic Research Project (Hunt, Lundberg and Zuckerman) will start living a life of their own.

The creation of digital research objects

For the reasons outlined in the preceding section, the creation of digital objects—be it images of inscriptions or manuscripts, electronic versions of ancient corpora, or collections of secondary literature—is a crucial part of humanities research. It is more than just preparation for research (cf. Talstra). This is a fundamental difference between databases as they are used in the humanities and those that are used in the natural sciences. The way in which inscriptions are photographed (cf. Hunt, Lundberg and Zuckerman) or in which text corpora are transcribed and encoded, is crucial for the way in which these research objects will be studied in the future. The creation of the digital objects has to meet the standards of the various disciplines involved, such as paleography, linguistics, and philology (cf. Talstra). Even in image

---

[26] This issue was raised by Ernst Thoutenhoofd in the closing paper to the colloquium.

capture and editing, which may at first sight be a rather straightforward and 'objective' procedure, 'virtually all parameters in the process (...) require intellectual, critical choices, interpretation, and manipulation' (Dahlström; see further Parker's section on 'transcription as interpretation').

The same applies to other processes of the creation of digital objects, such as linguistic encoding and textual editing, which are full of decision-taking and disambiguation. Annotation and encoding, for example, most often concern the enrichment of a text by information that is in the head of the human researcher. These activities give rise to new challenges and sometimes they make us think about things we had never thought about. If we have to take disambiguating decisions regarding, for example, the identification of participants or the assignment of semantic categories such as animate/inanimate, male/female, we have to face questions such as: are there ways to use the textual data for such disambiguations? Are there signals in the corpus that can be used for that? (cf. Talstra).[27]

In addition to those digital objects that are intended to be representations of 'real' research objects, there are also new, 'digitally-born' objects, which can be subjected to scholarly research. This creates new challenges as well. David Crystal shows that the phenomenon of multi-authored texts of the wiki-type, for example, are not only stylistically and pragmatically heterogeneous, but also 'disturb our sense of the physical identity of a text': They raise the question how we are 'to define the boundaries of a text which is ongoing' (cf. below, the section on 'Innovation').

Scholarly and scientific research

The spread of digital technology across philology, linguistics, and literary studies suggests that textual scholarship itself is taking on a more laboratory-like image. The ability to sort, quantify, reproduce, and report text through computation would seem to facilitate the exploration of text as another type of quantitative data (akin to protein structures or geographic features of the seabed). The possibility to test hypotheses upon the data introduces experimentation into

---

[27] Thus Joost Kircz at one of the colloquium discussions.

textual scholarship (cf. Talstra).[28] And the opportunity to present the research data—not only the research results—digitally in one way or another, gives humanity scholars the chance to 'open up the doors to their laboratory' (see below; compare the online availability of 737 pages with appendices of Salemans' PhD dissertation, mentioned in his note 1). This development requires also the use of more precise terminological distinctions between terms that have different meanings in the sciences but are often used indiscriminately by humanities scholars. Thus 'data', 'information', and 'knowledge' can no longer be confused (see below) and 'hypothesis' can no longer be used as merely a synonym for 'idea' or 'assumption'.

This does not mean, however, that developing the potential of digital technology in textual scholarship only gives it a more scientific character. It rather highlights text analysis and text interpretation as two increasingly separated sub-tasks in the study of texts. The implied dual nature of interpretation as the traditional, valued mode of *scholarly* text comparison, combined with an increasingly widespread reliance on digital text analysis as *scientific* mode of inquiry raises the question as to whether the reflexive concepts that are central to interpretation—individualism, subjectivity—are affected by the anonymised, normative assumptions implied by formal categorisations of text as digital data.

The contributions to this volume provide a number of examples in which textual analysis receives a more 'scientific' character, in which formalisation is used as a means to overcome the individualism and subjectivity that characterizes much philological research, or in which generalisation of analytical procedures and formal and systematic registration of the data help develop research strategies that allow the verification of its conclusions. The contributions in Part Two (Dahlström, Lavagnino, Salemans) show how this can take place in textual criticism and text editing and they elaborate in different ways on oppositions such as scholarly versus scientific; objective versus subjective; nomothetic versus idiographic; and mass versus critical digitisation. On the one hand they show how in the humanities systematic, corpusbased approaches are gaining ground and how insights and methods

---

[28] On the notion of 'experiment' in biblical textual scholarship see also Talstra and Dyk, 'The computer and biblical research' and Van Peursen, 'How to establish a verbal paradigm'.

from the sciences can lead to new directions in textual scholarship. (A telling example is the application of insights from cladistics in text geneaology in Ben Salemans' contribution.) On the other hand, each of these three authors agree that 'scholarly judgement and experience are still needed' (Lavagnino) and that the subjective or intersubjective human judgement is indispensable.

Continuation, acceleration, and innovation

Imitation of traditional practices or methodological innovation?

Many electronic instruments for textual scholarship that have become available over the last decades are imitations of traditional tools. Instruments for word searches in digital texts, for example, fulfil the same function as traditional concordances. Therefore, we invited the contributors to address the question as to whether the computer can do more than merely imitate, or speed up, traditional practices. Where does the use of the computer lead to new research strategies? Is there really methodological innovation? In other words, the question is 'whether we are simply going to speed up classical techniques of collecting and sorting data or whether we are really developing a new domain of techniques for scholarly access to classical texts (…) Can we get beyond just imitating the classical instruments for sorting lexical materials (concordance, lexicon) and for comparing parallel literary texts and fragments (synopsis)?' (cf. Talstra)[29] Is our use of the digital medium in fact an imitation of the medium of the book, or do we start to employ the salient characteristics of the new medium (cf. Van der Weel)?

Continuation and imitation

The question formulated in our invitation may erroneously have given the impression that the imitation and even the acceleration of classical techniques is of little or no value. But this is by no means the case. First, even the imitation of classical techniques that have crystallised over the centuries is an achievement in itself. This applies, for example, to the evolution of the computer from a calculating machine

---

[29] See also Dyk and Talstra, 'The computer and biblical research'.

to a language machine and its subsequent use for word processing and desktop publishing in the first stage of the history of computing.[30] Even though this use concerned mainly the imitation of certain aspects of conventional text production, 'there is no doubt that it was a triumph that we managed this replication of analogue textual practices and print functions in the digital realm so well' (Van der Weel).

Even things that seem easy for a human researcher are complex to implement with computer programs. This applies, for example, to the parallel alignment of texts in a synopsis. Eep Talstra argues that the imitation of what has been done before manually is the first step in the development of innovative computational research strategies.[31] The need to make explicit the parameters that are taken into account, is in many cases a step forward compared with traditional, intuitive approaches. The preparation of a synopsis involves a large variety of observations of lexical, syntactical, and literary correspondences, and decisions about the weight attached to them.[32] Most of these observations and decisions remain implicit if a synopsis is made manually by a human researcher. But they have to be made explicit if computer programs are instructed to establish corresponding elements in multiple texts. In this way the use of the computer forces us to sharpen our methodological focus and to work from explicit points of departure rather than implicit presuppositions.[33]

Also in text-genealogical research the need to make parameters and criteria explicit constitutes a step forward. This appears, for example, from Ben Salemans' contribution, in which he describes the need he had to formulate a list of characteristics of text-genealogical variants and to justify the acceptance or rejection of a variant as a building element for the text-genealogical chain on the basis of these characteristics.

---

[30] Cf. Van der Weel, *Changing our textual minds: Towards a digital order of knowledge* (forthcoming from Manchester University Press), Chapter 4: 'History of text and the computer'.

[31] See also Van Peursen and Talstra, 'Computer-assisted analysis of parallel texts.'

[32] Cf. Lasserre, *Les synopses.*

[33] In a pilot study on the biblical books of Kings and Chronicles the calculation of corresponding lexemes revealed some interesting parallels that were not represented in the existing printed synopses, where no justification for disregarding these parallels had been given. The synopses appear to have given priority to the question of which textual units tell the same episode rather than recognizing all the lexical correspondences; see Van Peursen and Talstra 'Computer-assisted analysis of parallel texts.'

Acceleration

A second modification to the question as we stated it in our invitation concerns acceleration. The question as to whether 'the computer can do more than merely imitate, or speed up, traditional practices' should not be read to imply a contrast between acceleration and innovation. Acceleration itself can lead not just to a quantitative, but also to a qualitative change. So even speeding up things is by itself innovative. The quantitative increase triggers a qualitative transformation, due to, for example, the enormous searchability. The same can be said of the new media for collaboration. At first sight they just facilitate networking and provide cheaper and faster, alternatives to letters sent by regular mail or international conferences. However, the processes of collaboration, which can be easier, faster, and more frequent thanks to the networked computer, allow cooperating scholars to discover patterns and regularities that would never be found by scholars working in isolation.

Innovation

Some contributions make us aware that sometimes what seems to be very innovative, such as the use of the calculating and sorting power of the computer, has its predecessors in works from long before the digital era. This applies even to one of the most salient features of electronic text, its hyperlinked nature. At a conceptual level 'hyperlinking', as the opposite of a linear, unidimensional representation of knowledge, relates to the models of thinking that we use to classify the world, and is related to the human condition rather than to computer technology.[34] At a practical level, non-linear representations of information can also be found before the digital era, as the diagrams in Jacob Lorhard's *Ogdoas Scholastica* at the beginning of the seventeenth century show (Sandborg-Petersen and Øhrstrøm). Even the footnote can be construed as a form of hyperlinking.

Technology to support the non-linear organisation of information has its predecessors in the pre-digital era as well. An interesting example is the bookwheel (see Plate 7), an instrument developed in

---

[34] For classification as an inherent element of the human condition and its role in information systems and modern information technology, as well as in history and society, see Bowker and Star, *Sorting things out*.

the Renaissance that could hold various books, and that enabled easy consultation of these books by just turning the wheel. The bookwheel could be used, for example, by lawyers, who consulted various sources on the same subject, or by biblical scholars, who could consult various commentaries on the same passage almost simultaneously.[35] Although going from one commentary to another by a click on the mouse may go faster than using the bookwheel, the instruments involved function basically the same.[36]

These observations do not deny, however, that IT does cause innovation. In some areas the new opportunities provided by technology are obvious. The application of computer science in the study of watermarks has made the invisible visible, and 'succeed[ed] where the naked eye and ad-hoc techniques would have failed' (Boyle and Hiary). But there are more ways in which IT leads to transformation and innovation.

In the first place, the use of the computer changes the basic concepts and categories of philology, including the notions of 'text'—both as a means of communication and as an object of transmission—'edition', and 'annotation' (see below, the section 'Transformations of concepts and categories').

In the second place, research objects themselves change. In the digital environment new types of texts are produced, such as blog posts, contributions to wiki's, emails and the like. These texts show peculiar new linguistic features that deserve linguistic analysis (Crystal); they are subjected to social agreements of various kinds and hence provide interesting material for a sociological analysis (Thoutenhoofd); and they are part of 'the explosion of cultural artifacts that are born digital', which should be taken seriously by humanities scholars if they 'wish to continue to study all of the humanities in their varied forms' (Zafrin). Further, in the digital world we see new ways in which groups are created and identity is defined, which can be subjected to virtual ethnography (Thoutenhoofd). As we have seen above, the

---

[35] Cf. Marcus, 'The silence of the archive and the noise of cyberspace', 19; Van der Weel, 'Explorations in the libroverse'; see also Sawday, 'Towards the Renaissance computer'.

[36] Cf. Marcus, 'The silence of the archive and the noise of cyberspace', 19: 'The computer is a bit less unwieldy. Online editions of early modern texts can function very much like such a scholar's wheel, only ampler and much more compact, allowing modern scholars to locate and correlate passages on similar subjects with great ease and precision.

digital representations of traditional research objects create research objects in their own right, which become more and more independent of the 'real' objects they represent. It is in this context that the inter-action between technological innovation and the 're-awakened desire for presence' mentioned above in the section 'Presence, meaning, and interpretation' becomes a pressing concern.

A third area in which innovation takes places is that of text editing and text comparison. Processes of text production and transmission have led to a multitude of parallel compositions.[37] (cf. Talstra) The traditional ways of dealing with these works in synopses or critical text editions are bound to the limitations of paper format and size. In the last decades we have seen a number of initiatives to use the computer for text representation and comparison. One example presented in this volume is the edition of *Ogdoas Scholastica*, which implements text as hypertext (Sandborg-Petersen and Øhrstrøm).

Ongoing developments

The question as to the innovation brought about by the computer has no final answer, given the swift changes in digital humanities. The three stages of 'the history of computing as a technology and as a digital textual medium' identified by Van der Weel, which can be character-ised by the catch words imitation, dissemination, and democratisation, cover only a few decades. They show a rapid process of innovation, invention, and discovery, rather than a dichotomy between the pre-digital and the digital era. Roberto Busa's digital text of the works of St Thomas Aquinas, one of the first large digital texts, is a milestone in the early history of digital humanities, but its accuracy left much to be desired (Lavagnino) and its approach was conventional rather than innovative (Van der Weel). Since then, however, tremendous progress has been made, both methodologically and technically. This applies not only to corpus creation, but also, for example, to image capturing. The results of recent infrared captures of ancient manuscripts com-

---

[37] On 'works' (rather than 'texts') that are subjected to ongoing processes of com-position and revision, see Voorbij, 'The *Chronicon* of Helinand of Froidmont', 5: 'The editorial team decided to produce a "historical-critical, genetic edition". This type of edition tries to give an understanding of the process of composition and textualization of the *Chronicon*. It is not "the" text of the *Chronicon*, nor even "a" text thereof that will be edited, but rather the *Chronicon* as a "work".'

pared with the results of two decades ago show that technologies 'have dramatically improved' (Hunt, Lundberg and Zuckerman).

Given the rapid changes we are witnessing, while being still at the treshold of a digital order, we can only imagine how the technical and methodological innovation will evolve in the future and what impact this will have on future textual scholarship. Digital technology is more than a response to challenges of traditional philology. It does not provide the final answer to questions raised in the past, nor does it promise to do so in the future. It rather becomes part of the framework of reference of research, raises new questions and helps develop new research strategies. This becomes clear, for example, in the way in which the texts that are studied and the digital instruments that are used in that study are interwoven in an electronic annotated text (see following section).

Transformations of concepts and categories

The innovation brought about by IT changes the basic concepts and categories of philology. The notion of 'text' as a means of communication changes due to 'digitally mediated communication', which differs both from speech (for example tone-voice substitutions) and from writing (for example hypertext linkage, framing) (Crystal). The notion of 'text' as object of transmission has changed due to the potential of electronic editing. Traditionally scholarly text editions contain a main text and a critical apparatus, and the establishing of the main text and the selection of variants are based on the text-critical research of the editor. However, the conceptualisation of the notions of 'text', 'variant', and 'edition' should be redefined in digital textual scholarship (Dahlström; Lavagnino).[38] Especially in the study of interrelated textual artefacts from before the invention of printing, the new models for the representation of textual variation are useful. Thus regarding biblical studies, Talstra comments that '[t]he main challenge comes from the fact that biblical texts are part of the large collections of classical texts that have been produced long before the invention of the art of printing (…) We only possess *individual copies* of a text, none of them being identical to any other existing copy.'

---

[38] See also Buzzetti, 'Digital representation and text model'; Eggert, 'Text-encoding, theories of the text, and the "Work-Site"'; Schmidt, 'Graphical editor for manuscripts'.

Other notions that require further discussion due to the new tools for knowledge creation include annotation, interpretation, hermeneutics, explanation, and commentary. Many new tools integrate text and annotation, so that they are not separate, but melt together. In the digital age, annotation is a completely new field, which includes not only traditional scholarly commentary, but also social tagging, blog comments, and comments solicited via specialised software (Zafrin). The transformation of concepts brought about by IT is also visible in more general philosophical and scientific terminology. This is illustrated by the transformations that the word 'ontology' underwent when it was imported from philosophy into computer science (Sandborg-Petersen and Øhrstrøm).

Knowledge creation and representation

Data, information, and knowledge creation

In the invitation to the colloquium, we gave the following description to the theme session 'Knowledge creation and representation':

> Text, as a record of ideas, a means to construct author(ity), and a material carrier of communication between humans has been central not only to philology, but to scholarship generally. Text is knowledge represented as matter: visible and revisitable, portable and measurable. As discipline focused on understanding texts and change in texts critically, philology is therefore a unique scholarly resource for understanding ways in which text alters under conditions of new technology, but of course knowledge of text in philology is itself also a text, both epistemologically specific and formally encoded (theorised). Does new technology make philological approaches and insights into the nature of text more transparent for other scholars? What broader challenges, shared interests and opportunities emerge as text comparison becomes part of a wider move towards integrated forms of (collaborative) e-research and the multi-purposing of data collections?

The innovation brought about by IT demands that also in the humanities clear distinctions are made between data, information, and knowledge. The movement to a more laboratory-like image of digital scholarship (see above, 'Scholarly and scientific research') requires from philologists that they start with the rough data, such as the sequences of graphemes that constitute a text. When the data are processed to be useful, and when meaningful connections between data are established, so that they provide answers to 'who', 'what', 'where', and 'when' questions, the data become information. The appropriate collection of information results in knowledge, which provides answers to the 'how' and 'why' questions.[39]

This process of knowledge creation raises a number of challenging questions. What is the role of the computer in the analysis of the data, the retrieval of information, and the creation of knowledge? Much data can now be made easily available, but how to develop abstract concepts for the preservation of 'concrete information'? (Hunt, Lundberg and Zuckerman) Does the digital medium lead to new ways of collaborative knowledge creation? And what role does the computer play in knowledge representation?

The contributions to this volume address various aspects of these questions. As noted in the preceding section, digitally mediated communication differs from speech as well as writing and leads to a new notion of text (Crystal). Hyperlinking, for example, can drastically change the representation of information (Sandborg-Petersen and Øhrstrøm). The internet creates new possibilities of worldwide data distribution, but it also provides new challenges regarding storing, distribution, and presentation of information (Hunt, Lundberg and Zuckerman), and new questions concerning responsibility and authorship (Crystal) and copyright (Schmid). New tools for collaboration and text annotation raise the question as to how we can overcome the individualism of humanities scholars (Zafrin). Interesting case studies include the tool by which various scholars annotate fourteenth-century Italian texts on the site of the Virtual Humanities Lab, Brown University (Zafrin) and the digital workflow used in the preparation of a text edition of the New Testament (Schmid), which employs the advantages of the electronic medium for collaboration.

Authority

The new means of knowledge creation and representation raise the question of authority. Two contrasting tendencies are at work. On the one hand, IT leads to more rigor and sharper definitions of analytical procedures due to the 'scientification' of the humanities; on the other hand, the new media lead to new roles and participants in knowledge creation that run counter to the strictness in the application of

---

[39] For the definitions of 'data', 'information' and 'knowledge', see www.systems-thinking.org/dikw/dikw.htm.

scientific principles. The authority of the text editor seems to vanish due to the creation of digital archives, by which the editors, rather than presenting an established critical text, 'open up the doors to their editorial "lab".' (Dahlström) Van der Weel calls this 'the deferral of the interpretive burden, which shift[s] more and more from the instigator of the (scholarly) communication to its recipient'. Likewise, the traditional authority of scholars whose expertise was incorporated in established reference works seems to be replaced by the multi-authored, anonymous, and 'democratic' Wikipedia articles (cf. Van der Weel on 'democratic forms of knowledge production' that characterise the third stage of the history of computing). There seems to be an analogy between the situation that we are in now and the Reformation in the sixteenth century. The Reformation involved a challenge of the existing authorities that was supported, if not enhanced by the new medium of the printed book. Now, again, at the beginning of the digital era, the introduction of a new medium supports or enhances challenges to existing scholarly authorities.

Although much can be said in support of the claim that a new revolution is on its way, a closer look at the things that are going on at the beginning of the digital era and the situation before the digital era, warn us not to exaggerate the changes brought about by IT. On the one hand, we see right now, in the beginning of the digital era that new regulations are installed, which give authority to, for example, moderators of the Wikipedia or discussion lists. In some cases the regulations were installed after a completely 'democratic' (or 'anarchistic') list was demolished by some fanatical non-experts. The Virtual Manuscripts Room, which is being developed in Birmingham and Münster, will in the future serve as a sort of wiki for 'accredited researchers' (Parker; cf. also Schmid). Apparently there is a desire for such a restriction even in the third stage of the history of computing, that of 'democratisation' (Van der Weel; see above).

Yet, also in the Order of the Book (see below) authority is not beyond challenge. Carotta's book *Jesus was Caesar. On the Julian origin of Christianity* (Soesterberg: Aspect, 2005) in which he claims that the historical Jesus was identical with Julius Caesar was—with due respect for Aspect—not published by an acknowledged academic publisher, it did not appear in a peer-reviewed series edited by university professors, and I do not know any New Testament specialist working in an established institution for Higher Education, who takes this book seriously. This seems to demonstrate how in the Order of the Book authority is well defined in terms of channels of publication. However, the book has been published anyway, and the struggle for authority, especially among the larger public, is not over. Carotta's book has raised more interest from the large public than many specialised studies by New Testament professors, and even a professor of Leiden University—from the Faculty of Law—revealed himself as a strong advocate of Carotta's thesis. Other examples, such as Dan Brown's *The Da Vinci Code* show that the authority of established academic scholars is not taken for granted in society, and that books that challenge or ignore scholarly 'authorities' may reach a wider readership and selling numbers of which every serious humanities scholar would be envious.

However interesting and innovative all the applications of the digital medium are, we have to be aware that we are still at the very beginning of the digital era and that, as Van der Weel puts it, we still live in 'the Order of the Book' rather than in a digital order. Van der Weel recalls that in the past new means of communication had a long way to go before their salient characteristics were fully employed or even recognised. Their usage started as an alternative to the familiar forms of communication: writing complementing speech, and print complementing writing. So the challenge is how we can avoid imitating the familiar medium and instead make full use of the salient characteristics of the unfamiliar new medium to discover the best possible use we can make of it for the task of creating 'knowledge instruments' such as thematically structured research collections.

This volume

As indicated in the beginning of this introduction, the contributions to the present volume were triggered by a KNAW colloquium, but the contents of the contributions have been modified and expanded. The present arrangement in four sections is also different, because it appeared impossible to pin-point each of the written contributions to one of the session themes of the colloquium. Moreover, we wanted to do justice to the fact that the character of the contributions differs in that some address more abstract, reflexive questions, whereas others present case studies that illustrate various aspects of the colloquium theme.

Section One, entitled 'Continuation and innovation in e-philology', starts with traditional, pre-digital research. In the first chapter Eep

Talstra shows that many practices in digital philology have their predecessors in the pre-digital era. Talstra addresses the question as to
whether we can imitate, or even go beyond, the traditional tools. Taking a more philosophical perspective, Ulrik Sandborg-Petersen and
Peter Øhrstrøm discuss in the second chapter the import of the word
'ontology' from philosophy into computer science. They trace the origin and history of this word and its development from something that
deals with 'the entities that exist, as well as their qualities and attributes', towards 'something more subjective and changeable', which
deals with 'constructing formalised, semantic, model-theoretical, and/
or logic-based models which can easily be implemented in computer
systems'.

Section Two, entitled 'Scholarly and scientific research', contains
contributions that address different modes of research expressed by
such oppositions as scholarly versus scientific; objective versus subjective; nomothetic versus idiographic; and mass versus critical digitisation. In Chapter 4 Mats Dahlström discusses two library digitisation
approaches: mass digitisation and critical digitisation. He addresses
the related question as to whether editions are presentations of facts
or hermeneutical documents and subjective interpretations. In Chapter 5 John Lavagnino presents a systematic approach to emendations
and describes the role that corpus methods can play to automate the
collection and evaluation of large amounts of relevant data. In Chapter 6 Ben Salemans shows how the computer can help implement an
intersubjective, repeatable and controllable theory. These contributions show the new, promising directions that textual scholarship can
take thanks to more 'scientific' digital approaches. At the same time
they demonstrate that the subjective and intersubjective judgment of
the human researcher is indispensable and that creativity is needed
in digital textual scholarship (cf. above, the section on 'Scholarly and
scientific research').

Section Three presents five case studies. They show how in concrete cases the computer can support philological research and stimulate the development of completely new research areas. The projects
presented in this part address various aspects of digital imaging, collaboration tools, and digital workflows, as well as the innovation that
computer science can bring about in the technical, material study of
textual artefacts. In Chapter 7 Roger Boyle and Hazem Hiary show
how computer science and artificial intelligence provide new opportunities for watermark location and identification. In Chapter 8 Leta

Hunt, Marilyn Lundberg, and Bruce Zuckerman discuss the challenges they faced in two closely related projects, the West Semitic Research Project and the InscriptiFact Digital Library. They discuss the theoretical model they developed for the conceptualisation of images of texts as 'concrete abstractions'. In Chapter 9 David Parker describes a project that includes digital imaging and an electronic transcription of the Codex Sinaiticus, one of the oldest manuscripts of the Bible, which is currently dispersed over four locations. Ulrich Schmid's contribution (Chapter 10) deals with the preparation of text editions of the New Testament. He describes a digital workflow that 'makes full use of the advantages of the electronic medium (collaborative, modular, updatable) while at the same time meeting the accepted standards of the printed book (sustainability, accessability, accountability)'. In Chapter 11 Vika Zafrin describes a collaborative text annotation engine that was created as part of the Virtual Humanities Lab at Brown University as well as some other commenting engines that have been used by humanities scholars for text editing and annotation.

Altogether, the projects presented by these authors provide a fascinating collection of examples of the use of new technologies, which can be of tremendous help in textual scholarship (Hunt, Lundberg and Zuckerman) and lead to the development of complete new research areas (Boyle and Hiary); the digitisation of analogue research objects and the subsequent development of the thus created digital research objects in their own right (Parker, Hunt, Lundberg and Zuckerman); new directions in knowledge creation and representation making use of collaboration tools and digital workflows (Schmid, Zafrin); and the continuity with scholarly practices of the pre-digital era (especially Parker and Schmid).

Lastly, Section Four provides three wider perspectives on the developments investigated in this volume. In Chapter 12 David Crystal approaches the rapid changes that have taken place over the last decades in digitally mediated communication from a linguistic perspective. In Chapter 13, Adriaan van der Weel compares the digital medium as 'knowledge instrument' with the printed book and explores the interrelatedness of technological inventions and social uses of the digital textual medium, especially concentrating on the social digital practices of humanities scholars. Finally, in Chapter 14 Ernst Thoutenhoofd provides a broader sociological perspective. He takes digital textual scholarship as part of a wider cluster of scientific and scholarly research activities that draws on new technologies in which expert knowledge production is no longer an exclusively human undertaking.

References

Alter, R., *The five books of Moses. A translation with commentary* (New York, 2004).

Benjamin, W., 'Das Kunstwerk im Zeitalter seiner technischen Reproduzierbarkeit' (Zweite Fassung), in *idem, Gesammelte Schriften* 7/1 (ed. by R. Tiedemann and H. Schweppenhäuser; Frankfurt am Main: Suhrkamp, 1989), 350–385.

——, *Illuminations* (ed. by H. Arendt; translated by H. Zohn; New York, 1969).

Bowker, G.C. and S.L. Star (eds), *Sorting things out. Classification and its consequences* (Cambridge, Mass/London, 1999).

Buzzetti, D., 'Digital representation and text model', *New literary history* 33 (2002), 61–88.

Eggert, P., 'Text-encoding, theories of the text, and the "work-site"', *Literary and linguistic computing* 20 (2005), 425–535.

Gumbrecht, H.U., *The production of presence. What meaning cannot convey* (Stanford, CA, 2004).

—— *Powers of philology. Dynamics of textual scholarship* (Urbana, 2003).

Lasserre, G., *Les synopses. Élaboration et usage* (Subsidia Biblica 19, Rome 1996).

Marcus, L.S., 'The silence of the archive and the noise of cyberspace', in *The Renaissance computer. Knowledge technology in the first age of print*, eds N. Rhodes and J. Sawday (London and New York, 2000), 18–28.

Peursen, W.Th. van, and E. Talstra, 'Computer-assisted analysis of parallel texts in the Bible. The case of 2 Kings xviii–xix and its parallels in Isaiah and Chronicles', *Vetus Testamentum* 57 (2007), 45–72.

Peursen, W.Th. van, 'How to establish a verbal paradigm on the basis of ancient Syriac manuscripts', Proceedings of the EACL 2009 workshop on computational approaches to Semitic languages, 31 March 2009, Megaron Athens International Conference Centre, Athens, Greece. Available online at http://www.aclweb.org/anthology/W/W09/W09-0800.pdf.

Reedijk, Wim, Zuiver lezen. De Bijbel gelezen op de wijze van de vroegchristelijke woestijnvaders (PhD dissertation Vrije Universiteit Amsterdam; Budel, 2006).

Sawday, J. 'Towards the Renaissance computer', in *The Renaissance computer. Knowledge technology in the first age of print*, eds N. Rhodes and J. Sawday (London and New York, 2000), 29–44.

Schmidt, D., 'Graphical editor for manuscripts', *Literary and linguistic computing* 21 (2006), 341–351.

Steiner, George, *Real presences. Is there anything in what we say?* (London: Faber & Faber, 1989).

Talstra, E., and J.W. Dyk, 'The computer and biblical research. Are there perspectives beyond the imitation of classical instruments?' in *Text, translation, and tradition*, eds W.Th. van Peursen and R.B. ter Haar Romeny (Monographs of the Peshitta Institute Leiden 14. Leiden, 2006), 189–203.

Voorbij, H., 'The *Chronicon* of Helinand of Froidmont. A printed edition in an electronic environment', in *Produktion und Kontext*, ed. H.T.M. van Vliet (Beihefte zu *Editio* 13; Tübingen, 1999), 3–12.

Weel, A.H. van der, 'Explorations in the libroverse', lecture given at the 147th Nobel Symposium, 'Going digital: Evolutionary and revolutionary aspects of digitization', Stockholm, Royal Academy of Science, 23–26 June 2009; available online at http://www.let.leidenuniv.nl/wgbw/research/Weel_Articles/NobelPaperPub.pdf.

——, Changing our textual mind: Towards a digital order of knowledge (forthcoming); available online at http://www.vanderweel.demon.nl/Book/Index.html.

Wise, M.O., *A critical study of the Temple Scroll from Qumran Cave 11* (Studies in Ancient Oriental Civilization 49; Chicago/Illinois, 1990).

PART ONE

CONTINUATION AND INNOVATION IN E-PHILOLOGY

IN THE BEGINNING, WHEN MAKING COPIES USED TO BE AN ART…THE BIBLE AMONG POETS AND ENGINEERS

Eep Talstra

What is it that biblical scholars want to know about the ancient texts they study? What are the instruments they use for their analytical purposes? As soon as one begins to look at pictures of Dead Sea scrolls on the internet, or to admire the wealth of illuminated medieval manuscripts presented there, one becomes aware of the complexities of the seemingly simple traditional task of the philologist: how to get access to these texts, not as pictures, but as texts. The traditional scholarly tasks have always been relatively simple to formulate: we need to make these texts accessible to scholars. That means making the manuscripts available on our desk as scholarly editions. Can we make them accessible to a broader audience? Then we should not only learn the ancient languages, but also produce a translation of the manuscripts. Can we find out more about the conditions of their production, their author and their intended readers? Then we need to compare our text and its language with related documents from a similar cultural background.

These simple goals have always defined the complex tools to be used. For making textual editions we need the discipline of textual criticism, that is, the comparison of different readings in various manuscripts. For making translations we need grammars and lexicons. For the research in textual composition we need instruments that allow us to compare various literary texts, such as concordances (presenting words in context) and synopses (presenting parallel texts).[1] The intriguing question now is whether the production and the use of these instruments can be imitated by the use of computers. Or, even better, can we go beyond imitation? Can computer-assisted research of ancient texts provide us with better and more flexible instruments to achieve the goals set by the classical discipline of philology?

---

[1] K.D. Jenner et al., 'An interdisciplinary debate'.

Bible and computing. A new tool, and a new goal?

On methods and tools

From the very beginning of the introduction of computers into the area of biblical studies (before 1970)[2] the question has been asked whether we are simply going to speed up classical techniques of collecting and sorting data or whether we are really developing a new domain of techniques for scholarly access to classical texts. The main challenge comes from the fact that biblical texts are part of the large collections of classical texts that have been produced long before the invention of the art of printing. These texts are from a cultural period that did not have access to modern techniques for the multiplication of texts, neither electronic ones nor mechanical ones. This means that the material status of the text has been ambiguous for a long time, since we only possess *individual copies* of a text, none of them being identical to any other existing copy. So, what is 'the' text of a classical literary work?

Only when one is aware of this particular status of ancient texts and their traditions can one ask in a meaningful way the question of how to use computer technology in the domain of Bible and philology.[3] So far specialists in the area of Bible and computing have been experimenting with *three options*. They have done so by following closely, not so much the order of possible research questions in their own discipline, but rather the order of the hardware and software being developed. This means that the following three stages may be discerned.

a. In the era of mainframe computers, line printers and punched cards the main question addressed was: can the computer imitate the *scholar-philologist*? For the goal of textual criticism one should be able to produce electronic copies from classical texts, such as the Hebrew Bible, the Old Testament in Greek or the New Testament, based on the main manuscripts or on the existing scholarly editions. These electronic texts need to be linguistically analysed minimally at word level in order to be able to study and compare them.[4] The

---

[2] G.E. Weil and F. Chénique, 'Prolegomènes à l'utilisation de méthodes de statistique linguistique'.

[3] E. Talstra, 'Desk and discipline'.

[4] E. Tov, *A computerized database for Septuagint studies*.

computer should assist us with the production of statistics of variant readings. So in this first period, from the seventies, computers were used to produce mainly concordances and to produce some synopses of particular linguistic data. This was an important start: the computer proved to be of assistance in the basic task of philology, that is, the comparison of manuscripts and texts. However, these interests led to a next step: can we proceed from electronic copies of a manuscript to the linguistic analysis of the literary text of that manuscript?

b. Already in this early period of mainframe computers one can find a number of projects that are based on the next question of whether the computer can imitate the *scholar-reader*. Can we get beyond just imitating the *classical instruments* for sorting lexical materials (concordance, lexicon) and for comparing parallel literary texts and fragments (synopsis)?[5] In other words, can we proceed from comparing manuscripts to the area of the study of languages? Thus as early as the seventies substantial projects were started with the goal of having electronic copies of the texts of the Hebrew Bible, the New Testament or other classical texts with a morphological encoding, that is, a word-level grammatical analysis (for example The Computer Bible, CATSS, Gramcord; my own research group also started in that period: 1977).[6] Text databases were built to allow for linguistic research, so as to be able to search for and find sets of particular expressions from Hebrew idiom, Greek syntax, and so on. However, one can also observe that due to the technical limitations of that period many of the ideals of computer-assisted linguistic analysis of ancient texts failed to be realised. Moreover, the development of the internet, the PC and the enormous increase in the capacity of storing and transmitting data made philologists change their mind. If it is too demanding to make the computer imitate our analytical work, then the computer should allow us easy access to our data, the ancient manuscripts, in the form of pictures and electronic copies.

c. So, since about 1985 the research question changed again: Should the computer be used to imitate the *scholar-librarian*? Since the

---

[5] E. Talstra, 'On scrolls and screens'.
[6] A good overview of the research projects in this periode (1965–1985) one finds in the handbook compiled by J.J. Hughes, *Bits, bytes and biblical studies*.

introduction of the internet and the possibilities of graphical display this has become the most popular variant of using computers
in the humanities. The internet allows easy access to many manuscripts and text traditions by presenting all kinds of copies and
pictures of manuscripts on one's screen. This makes the computer
actually an extension of one's library and of the material present on
one's desk.

This development, however, has created a paradoxical situation. With a
growing capacity to store texts, with more speed available to allow easy
access to pictures and textual data, the use of computers to accomplish
the analysis of textual materials for philologists actually decreased in
its level of sophistication. Once a source text has been downloaded
from the internet, the major part of linguistic analysis in most cases is
still being performed as in classical times, that is to say in the mind of
the scholar.[7] The goal of giving access to *documents* could be achieved,
but the basic question remained: what about computer-assisted study
of the *texts* of the manuscripts?

Back to basics

Happily, times are changing once more. The colloquium on e-philology
of which the papers are included in this volume, and the very existence
of one of its organising institutions, the Virtual Knowledge Studio for
the Humanities and Social Sciences, are witnesses to that. In the last
ten years the earlier research question (b) on method and textual analysis in the domain of philology and humanities has been addressed
again: can the computer imitate the *scholar-reader*? Recently, Patricia
Alkhoven and Peter Doorn presented a short introduction to these
new developments in the humanities in general in their 'New research
perspectives for the humanities'.[8]

In my contribution, therefore, I will concentrate on the question
concerning what it means when biblical scholars in our days again try
to enter the area of computer-assisted textual research. Traditionally

---

[7] See E. Talstra, 'The Hebrew Bible and the computer', which appeared in volume
1 of *International journal of humanities and arts computing*. (In this journal one finds
a number of the contributions to the 3rd SURF symposium 'ICT en Humaniora',
September 30–October 1, Amsterdam 2004.)

[8] P. Alkhoven and P. Doorn, 'New research perspectives for the humanities'.

the biblical scholar has to be aware of three different layers of information in an ancient text:

> A text as a *literary composition transmitted in various manuscripts*;
> A text as a *linguistically organized discourse structure*;
> A text as a *source for the study of an ancient language*.

My main statement in this contribution will be that we will only enter a new research mode here if we are capable of finding systems and procedures that allow for various ways of access to these three different layers of the ancient texts: (1) as artefacts of literary art, created, adapted, and transmitted by many generations; (2) as products of linguistic communication in an ancient context; and (3) as part of a text corpus, witness to an ancient language. To illustrate this I present an example that is well known to biblical scholars: the transition from Isaiah Chapter 63 to Chapter 64 in the Hebrew Bible. Here one can observe that the tradition of textual transmission does not fit the text's linguistic structure. I present the text first without any linguistic subdivision, following the translation of the New Revised Standard Version. These chapters express a complaint by the early Judean community who, living in difficult circumstances (probably the Hellenistic period), describe their feelings of being left alone by the LORD (YHWH).

63:19 *We have long been like those whom you do not rule, like those not called by your name. O that you would tear open the heavens and come down, so that the mountains would quake at your presence*

64:01 *as when fire kindles brushwood and the fire causes water to boil—to make your name known to your adversaries, so that the nations might tremble at your presence.*

The mismatch between the structural division in the literary tradition and the linguistic division is obvious: the long sentence starting in the last verse of Chapter 63 is continued in Chapter 64, verse 1. The question arises as to how to analyse and store both the literary tradition and the linguistic structure in a text database. (I leave aside the fact that the chapter division is of a later origin than the Hebrew text itself.) We have to deal with the fact that we as philologists need a text database of the text as we have it now: including, not excluding its literary and document history.

The long manuscript tradition of the Hebrew text divides it into the categories of chapters, verses and half verses, and with even further subdivisions, all being marked by the accent signs put into the text. Official scholarly editions of the text will follow this manuscript tradition, respecting the divisions into chapters, verses and half verses. With poetic or prophetic texts, however, they will usually present a further subdivision of the texts into poetic lines, guided by the accent signs in the Hebrew text. Then it looks like this (the elements that have been added to the text in the NRSV translation I have placed within brackets):

63:19a   *We have long been* (like those) *whom you do not rule,*
                                                                          (zaqef)
           (like those) *not called by your name.*   (atnach [*half verse*])
63:19b   *O that you would tear open the heavens and come down,*
                                                                          (zaqef)
           (so that) *the mountains would quake at your presence.*
                                                          (sof pasuq [*end of verse*])
64:01a   *as* (when) *fire kindles brushwood*   (rebia)
           (and the) *fire causes water to boil—*   (zaqef)
           to make your name known to your adversaries,
                                                               (atnach [*half verse*])
64:01b   (so that the) *nations might tremble at your presence.*
                                                          (sof pasuq [*end of verse*])

For the sake of convenience I have given the names of the accents between brackets behind each line. A verse ends where one finds the *sof pasuq.* The verse is divided into two parts by the *atnach.* The half verses can be divided again into two parts by the *zaqef,* each of which may be subdivided further by the *rebia.* This procedure results in strings of words which, according to the tradition, should be taken together semantically. These strings, however, do not necessarily represent clearly defined syntactic units. The divisions reflect literary tradition. They help to recite the prophetic text and to interpret it when reading it. Thus the system of verse division based on the accents in the text has the advantage of representing a very old tradition of reading and interpretation, but it has the disadvantage of being primarily a literary interpretation and not a linguistic analysis.

In terms of syntactical analysis, therefore, one would need to present the text differently again, in order to create a system of one clause (predication) written on one line.

63:19A *We have long been* (like those)
        whom you do not rule,
        (like those) *not called by your name.*
63:19B *O that you would tear open the heavens*
        and come down,
        (so that) *the mountains would quake at your presence*
64:01A *– as* (when) *fire kindles brushwood*
        (and the) *fire causes water to boil—*
        to make your name known to your adversaries,
64:01B (so that the) *nations might tremble at your presence.*

This example may clarify where the first challenge is to be met, once we decide to analyse an ancient text in order to store it in a text database suited for linguistic and literary research.[9]

The three layers present in the text are basic material for three ways of studying texts from Antiquity. Computer-assisted textual analysis should be helpful to all three of them.

Layer 1, the text as a *literary composition*, that is, the literary form transmitted by one or more manuscripts and by ancient translations. This regards the text as a cultural artefact, part of human communication in a particular historical context: transmitted in manuscripts and by traditions of interpretation. We need to keep the traditional markers used in the text: chapters, verses, and accents, even when they run counter a strictly linguistic analysis. This is even more relevant when modern translations feel forced to overrule the chapter division, as is the case in this example, where the first lines of Chapter 64 are a direct grammatical continuation of the last lines of Chapter 63.

Layer 2, the text as a *discourse structure*. This is the linguistic, syntactic structure that one has to establish in order to be able to produce a grammatical interpretation and a translation of the full text. A composition is to be understood on the basis of linguistic knowledge, to be derived from the entire ancient text corpus (layer 3), in a way that is as independent as possible from the traditions of transmission and interpretation of a particular text. What is the discourse structure and how can we track the participants active in this text? Linguistic analysis of the text also requires that one leaves out the interpretative extras such as the ones used in NRSV: 'like those', 'so that', and so on.

---

[9] Chr. Hardmeier et al., *SESB: Stuttgart Electronic Study Bible.*

Layer 3, the text as a *source for the study of language*. The study
of the linguistic data present in the text that one is reading implies a
move from the individual composition to a general approach of the
language used in the entire literary corpus. Linguistic analysis is done
in order to develop systematic knowledge of classical Hebrew, Aramaic, Greek, and so on. It regards topics such as syntactic analysis,
analysis of clause structure, and verbal valency. For example: are the
interpretative extras as used in the NRSV: 'like those', 'then', and so on
correct in terms of syntax? Can they be applied in a consistent way?
Can one test them through comparison with other poetic texts?

This is what philologists using computer-assisted analysis want to
do: accept the presence of the three layers of information in ancient
literary texts and accept the fact that the study of one layer is interde-
pendent on the other layers. Literary composition and manuscript tradition tend to complicate, or even disturb the study of textual structure
and linguistic analysis. In the following section I will try to illustrate
this complicated relationship of computer-assisted textual analysis and
the tasks of philology. I will do so by a presentation of three examples
of texts taken from the prophets: Isaiah, Jeremiah, Ezekiel.[10] The goal
is also to present some experiments with description and interpretation, especially in the areas where linguistic and literary study meet or
even clash.

Experiment 1: Text and document tradition

Literary tradition creates or adapts a text into a structure that often
does not match patterns of grammar and syntax. So it generates the
problem of multiple hierarchies for data. The example here is from
Jeremiah 33: its verse division cuts one linguistic clause (with a number of embeddings and quotations) into two halves. The question is
whether one can do both: store or present a text in its traditional literary format and still be able to produce a concordance or a syntactic

---

[10] The tradition of the discipline appears to be that the study of Hebrew syntax
concentrates mainly on narrative texts. Poetry is left to rhetorical or stylistic analysis.
In our actual research (funded by the Netherlands Organisation for Scientific Research
[NWO]), called System and Design, we try to do syntactic analysis of poetic and prophetic texts.

parsing of clauses, based on correct linguistic units, not on literary segmentations.

Experiment 2: Text and discourse structure

The example chosen is from Isaiah 61, where one observes several unexpected changes of sender and addressee in the text: shifts of person, number, and gender. The challenge is how to keep track of the text's participants. Is it realistic to try to produce one linguistic structure of the chapter? Or should one leave it as it seems to be: a rather poorly coordinated literary text, showing traces of redactional activity that disturb its cohesion? The question is how far one can continue linguistic analysis of such texts. Literary tradition has generated a mode of literary interpretation that influences grammar and lexicon. Can computer-assisted research help us to generate more accurate grammatical knowledge of ancient languages in the area of syntactic structure and the role played in it by verbal tenses and by the rather frequent shifts of person and number?

Experiment 3: Text and multiple authorship

Literary tradition presents us with texts that have been composed, edited and recomposed. Biblical texts are as complicated as ancient buildings: they may have started as the product of someone's design, but during many generations of usage they have been changed, enlarged, repaired and renovated. The example chosen is Ezekiel 36 because of the complexity caused by repeated embedding of direct speech sections.[11] How does one analyse and store them as data, even when it is probable that this complexity results from redactional reworking of the text? The challenge for discourse research is how to keep track of the participants acting in the text. How does one define the linguistic domains of who speaks to whom and whose speech has been embedded in someone else's speech? In my view the general challenge in the area of e-philology is the tension or even the clash between the special

---

[11] For the complexities of direct speech sections in the book of Ezekiel see, S.A. Meier, *Speaking of speaking*, 230–242.

history of unique literary artefacts and the general processes of human cognition active in the actual process of reading the texts.

Linguistic structure and literary tradition: a strategy of
data production

Since literary texts from Antiquity, such as the texts from the Hebrew Bible, often have a particular poetic structure and also have been transmitted over centuries, one cannot expect the poetic structure, its manuscript tradition, and the linguistic structure made by syntactic analysis to match fully. Poetry exploits the lexicon or plays with the linguistic system; manuscript tradition adds further elements that represent continuing scribal activity, exegetical interpretation and also the effects of errors. The result of this complex process of textual production is the presence of multiple hierarchies. Any strategy to design a database for e-philology should find a solution to that challenge.

The strategy in linguistic analysis and data production we have adopted in our research group is, *first,* to avoid any mixture of linguistic data and literary interpretation in a text database; *second,* to give priority to linguistic analysis and *third,* to follow a procedure that analyses from form to function.[12] For example: the Hebrew verbal forms are not encoded in terms of past or present, completed or incomplete action, iterative or durative, but in terms of the morphological features and the grammatical functions they mark, for example, person, number, gender, conjugation. In a similar way the clause connections in a text are encoded by registering the markers present in the clauses: the conjunction and the verbal forms used, word order and possible lexical parallels. Further definition in terms of causal, consecutive, final, or another type of syntactic function has been postponed. It is the database itself that will be used to try to establish and implement them. In the first place, therefore, text syntactic structures are encoded in an interactive procedure, based on proposals calculated by a computer program from morphological, lexical, and clause type data. This results in a preliminary type of textual hierarchy, stored in a database that is not to be regarded as the final result of research into linguistic *functions*. Rather it should allow for further research into those functions.

---

[12] E. Talstra and C. Sikkel, 'Genese und Kategorienentwicklung der WIVU-Datenbank'.

Some examples of the initial encoding of clause connections used in the text segments presented below, are:

<312>   Conjunction *'we'* and *'yiqtol'* in second clause, *'Qatal'* in first clause.
<511>   Conjunction *'ki'* and *'yiqtol'* verbal forms in both connected clauses.
<200>   Formal identity of two clauses: identical conjunctions (if present), identical negations or modifiers (if present), identical verbal forms.
<201>   The same as <200> + coordinating conjunction in second clause.
<999>   Start direct speech section.
<222>   Clause parts interrupted by embedding.

Once the relations between (parts of) clauses have been calculated, one can use these codes as data for further calculation and encoding of syntactic features at a functional level. I will demonstrate this further below, in the next section on Isaiah 61.

Each clause in the database is also labelled according to the literary tradition: name of the book, number of the chapter, number of the verse. One needs to adhere to the literary tradition, otherwise the database would lose the commonly used system of textual reference. Here the complication of the multiple hierarchies becomes visible. The lines in Jeremiah 33:10b, 10g and 11a together build one clause, although interrupted by various types of embedding and quotations. The problem is the fact that this clause crosses the borders set by literary tradition: it runs from verse 10 into verse 11. Thus linguistic analysis and manuscript tradition do not match. For now we have decided to accept the hierarchy in the text marked by its literary and manuscript traditions as the system of reference, so the problem is that we have to do the linguistic definition of clauses only inside the literary boundaries and not outside them. The advantage is that results of queries and grammatical analysis can still be presented according to the document structure. The disadvantage is that one needs additional regulations where one needs to present, for example, an overview of syntactically correct clauses.

Experiment 1. Jeremiah 33:10–11: syntactic hierarchy and document structure

Syntactic structure	manuscript	Hebrew surface text in transliteration
| | | |-----|-----------------	JER 33,10a	[KH <Mo>] [>MR <Pr>] [JHWH <Su>]
| <200><999>		*–Thus | has said | YHWH*
| | |----|----|--------	JER 33,10b	[<WD <Mo>] [JCM< <Pr>] [B--MQWM H-ZH <Lo>]
| | <511><222> <16>		*"Again|will be heard|in this place*
| | | | |---|----	JER 33,10c	[>CR <Re>] [>TM <Su>] [>MRJM <PC>]
| | | | <999>		*of which | you (plur.) | are saying:*
| | | | |----	JER 33,10d	[XRB <PC>] [HW> <Su>]
| | | | <100>		*"A waste | it (is)*
| | | | |----	JER 33,10e	[M->JN <Ng>] [>DM <PC>]
| | | | <201>		*without | human being*
| | | | |----	JER 33,10f	[W-<Cj>] [M->JN <Ng>] [BHMH <Su>]
| | | |		*and | without | cattle"*
| | | |--------|----	JER 33,10g	[B-<RJ JHWDH W-B-XYWT JRWCLM <Lo>]
| | | <101> < 10>		*in the cities of Judah and in the streets of Jerusalem*
| | | | |----	JER 33,10h	[H-<Re>] [NCMWT <PC>]
| | | | <106>		*that | are desolate*
| | | | |----	JER 33,10i	[M->JN <Ng>] [>DM <PC>]
| | | | <201>		*without | human being*
| | | | |----	JER 33,10j	[W-<Cj>] [M->JN <Ng>] [JWCB <PC>]
| | | | <200>		*and | without | inhabitant*
| | | | |----	JER 33,10k	[W-<Cj>] [M->JN <Ng>] [BHMH <PC>]
| | | |		*and | without | cattle*
| | |---------------	JER 33,11a	[QWL FFWN W-QWL FMXH <Su>] [QWL XTN W-QWL KLH <Su>] [QWL >MRJM <Su>]
. . . . .		
. . . . .		
| | | <160> <999>		*the voice of joy and the voice of gladness,*
. . . . .		*the voice of the groom and the voice of the bride,*
. . . . .		*the voice of those saying:*
| | | | |----	JER 33,11b	[HWDW <Pr>] [>T JHWH YB>WT <Ob>]
| | | | <503>		*"Give thanks YHWH of Hosts*

Syntactic structure	manuscript	Hebrew surface text in transliteration
| | **|** | |----	JER 33,11c	[KJ <Cj>] [VWB <PC>] [JHWH <Su>]
| | **|** | <500>		*for | good (is) | Y*ʜᴡʜ
| | **|** | |----	JER 33,11d	[KJ <Cj>] [L-<WLM <PC>] [XSDW <Su>]
| | **|** |		*for | forever (is) | his mercy"*
| | **|** |------------	JER 33,11e	[MB>JM <PC>] [TWDH <Ob>] [BJT JHWH <Co>]
| | **|**		*bringing | thank offering | to the house of Y*ʜᴡʜ
| | **|------------------**	JER 33,11f	[KJ <Cj>] [>CJB <Pr>] [>T CBWT H->RY <Ob>] [K-B- >CNH <Aj>]
. . . | | <121>		*For | I will turn | the fortunes of the land | as at first'*
| | |------------------	JER 33,11g	[>MR <Pr>] [JHWH <Su>] *has said | Y*ʜᴡʜ
| |-----|------------------	JER 33,12a	[KH <Mo>] [>MR <Pr>] [JHWH YB>WT <Su>] *Thus | has said | Y*ʜᴡʜ:

The first line of verse 11 actually is a part of the same linguistic clause as in 10b and 10g. Verses 11a and 10g, therefore, should have been be encoded as being parts of one clause, interrupted by embeddings (i.e. by <222>), but since that conflicts with the literary tradition of segmentation (chapters and verses), it could not be done that way. So in this analysis verse 11a has been encoded <101> as if it were an elliptic clause asyndetically connected to 10g. For most of the search and retrieval actions this is not a problem. Only if one wanted an overview of all complete linguistic clauses of the chapter, or a concordance presenting lexemes in the context of their clauses, would one need extra measures to override the transition of the verse boundary.

Experiment 2. Isaiah 61: Discourse structure and tradition of interpretation

The experiment with this chapter represents a next step in the analytical procedures. It is an attempt to propose functional labels for the various clause connections that were established based on surface structure features, such as conjunction, verbal forms, clause type, and lexical parallels. The next procedure is to have a program read this preliminary structural analysis (as the one presented above with Jeremiah 33) and then let it propose a syntactic function to each couple of clauses established in the textual hierarchy by assigning additional labels to the clause connection, for example 'and', 'in order that', 'to do', 'while', 'because', and so on. This procedure is experimental in the sense that it tries to find first whether there are default functions that can be matched with the numerical codes calculated before. For example: can the code <511> (conjunction *ki*), verbal clause with *yiqtol*, connected to a preceding verbal clause with *yiqtol*, as in Jeremiah 33:11f and 10b, always be interpreted as 'X will be *for* Y will be'? And if not, what are the other features of the clauses one would have to analyse: a change of subject or complement, a shift of person, number, gender, different word order, semantic features of the verb?

This procedure begins as a search process for parameters that could be decisive for constructing various types of clause connections, rather than just being a procedure for direct text-level parsing. Once the insight in the linguistic parameters active on text and discourse level grows, the capacities of this procedure for actual parsing grow too. Thus the main goal is to answer the question how far one can get in analyzing syntactic structure in terms of a *consistent* grammar and syntax. Can one find a *system* (i.e. propose a syntax of clause connections) based on features of morphology, verbal tenses, conjunctions, and clause types?

Clearly this is an iterative procedure: it started with the interactive analysis of clause connections and their preliminary encoding by the numerical codes based only on surface data, as presented above. The next step is the re-analysis of these codes, testing them by proposing a linguistic interpretation for each of them. Difficulties encountered in the assignment of a functional interpretation may mean that either the functional analysis needs to take into consideration more clause features, or the clause connections that had been established in the first round are incorrect, or one is reaching a border line where semantic issues or pragmatic patterns of communication dominate the clause connections. Of course the difficulties encountered are the most exciting part of the procedure.[13]

---

[13] E. Talstra and C.H.J. van der Merwe, 'Analysis, retrieval and the demand for more data'.

The example of Isaiah 61 discussed in this section demonstrates some of the complexities that require additional experiments with syntactic analysis. For example, one finds:

- Unexpected *changes* of sender and addressee in the text: shifts of person, number, gender. In line 20ff. (verse 5) we find a transition from 'they' into 'you' (plural). Is that the same participant or a different one? How does one keep track of participants? Is it realistic to produce one linguistic structure of the text? Or should we leave it as it seems to be: a rather poorly coordinated literary text, showing traces of redactional activity?
- Rather uncommon cases of the order of *verbal tenses*. The *Weqatal* clause of line 14 has to be taken, as I will argue further below, as connected to the infinitive construction of line 10. How does one interpret this clause connection? Modern translations seem to avoid a decision when they simply start a new sentence: 'They will be called...'

Isaiah 61

Ln	Sent.Cl/	Cl.type	Ch.Verse	Text of NRSV translation	Syntactic function proposed
1	S: 1:1	NmCl	61,01	—The spirit of the Lord is upon me	
				| | | | | |<<120>>	יען =because JHWH has: משח =anoint
2	S: 2:1	xQtl	61,01 . | | _	because the Lord has anointed me	
				| | | |<<adjunct >><< 64>>	ל =to: בשׂר =bring good news
3	S: 2:2	InfC	61,01 . | | . _	to bring good news to the oppressed	
				| | | |<<120>>	0 he has: שׁלח =send objSfx 2= me
4	S: 3:1	0Qtl	61,01 . | _	He has sent me	
				| | | |<<adjunct >><< 64>>	ל =to: חבשׁ =bind up

(*cont.*)

Ln	Sent.Cl/	Cl.type	Ch.Verse	Text of NRSV translation	Syntactic function proposed

5　S: 3:2　InfC　61,01 . | . _　　to bind up the brokenhearted

　　　　　　　　　　　　|　|
　　　　　　　　　　　　|　|<<paralDep>><<200>>
　　　　　　　　　　　　　　--> 64　　　　　　　　קרא ל =to:
　　　　　　　　　　　　　　　　　　　　　　　　　　=proclaim

6　S: 3:3　InfC　61,01 . | . . _　to proclaim liberty
　　　　　　　　　　　　　　　　　to the captives

　　　　　　　　　　　　|　|　|
　　　　　　　　　　　　|　|　|<<ellipsis>>　　　　1 =and (to ..)

7　S: 3:4　Ellp　61,01 . | . . . |　_　and release to the prisoners

　　　　　　　　　　　　|
　　　　　　　　　　　　|<<paralDep>><<200>>
　　　　　　　　　　　　> --> 64　　　　　　　　קרא ל=to:
　　　　　　　　　　　　　　　　　　　　　　　　=proclaim

8　S: 3:5　InfC　61,02 . | . . . _　to proclaim the year of the Lord's favor and
　　　　　　　　　　　　　　　　　the day of the vengeance of our Lord

　　　　　　　　　　　　|　|
　　　　　　　　　　　　|<<paralDep>><<200>> >> --> 64　נחם ל =to:
　　　　　　　　　　　　　　　　　　　　　　　　　　=comfort

9　S: 3:6　InfC　61,02 . | . . . . _　to comfort all who mourn

　　　　　　　　　　　　|　|
　　　　　　　　　　　　|<<paralDep>><<200>> >>>　שׂים ל =to:
　　　　　　　　　　　　--> 64　　　　　　　　　　　=establish

10　S: 3:7　InfC　61,03 . | . . . . . _　to provide for those who mourn in Zion

　　　　　　　　　　　　|　　　　　|　|
　　　　　　　　　　　　|　　　　　|<<paralDep>><<200>> >>>>
　　　　　　　　　　　　　　--> 64　　　　　　נתן ל =to: =give

11　S: 3:8　InfC　61,03 . | . . . . . . |　_　to give them a garland in stead of
　　　　　　　　　　　　　　　　　　　　　　ashes

　　　　　　　　　　　　|　　　　　|　|
　　　　　　　　　　　　|　　　　　|<<ellipsis>>　　　0

12　S: 3:9　Ellp　61,03 . | . . . . . . | . _　the oil of gladness instead of
　　　　　　　　　　　　　　　　　　　　　　mourning

　　　　　　　　　　　　|　　　　　|　|
　　　　　　　　　　　　|　　　　　|<<ellipsis>> --> 104　　0

13　S: 3:10　Ellp　61,03 . | . . . . . . | . . _　the mantle of praise in stead of a
　　　　　　　　　　　　　　　　　　　　　　　faint spirit

　　　　　　　　　　　　|　　　　|
　　　　　　　　　　　　|<<324>>　　　　　　1 =then will be:
　　　　　　　　　　　　　　　　　　　　　　קרא =proclaim

14　S: 4:1　WQtl　61,03 . | . . . . . . _　They will be called oaks of righteousness,
　　　　　　　　　　　　　　　　　　　　the planting of the Lord

　　　　　　　　　　　　|　　　　　　|　|

(cont.)

Ln Sent.Cl/ Cl.type	Ch.Verse	Text of NRSV translation	Syntactic function proposed
		| |<<adjunct >><< 64>>	ל =to: פאר =beautify
15 S: 4:2 InfC	61,03	_ to display his glory	
		|<<322>> --> 324	ו=and (they) will: בנה =build
16 S: 5:1 WQtl	61,04	_ They shall build up the ancient ruins	
		| | |<<112>>	(they) will: קום =raise up
17 S: 6:1 xYqt	61,04	_ they shall raise up the former devastations	
		| |<<200>> -->322 -->324	ו =and (they) will: חדש =renew
18 S: 7:1 WQtl	61,04	_ they shall repair the ruined cities	
		| |<<ellipsis>>	0
19 S: 7:2 Ellp	61,04	_ the devastations of many generations	
		|<<322>> --> 324	ו =and זר =stranger [3pM] will: עמד =stand
20 S: 8:1 WQtl	61,05	_ Strangers shall stand	
		| | |<<200>> -->322 -->324	ו =and (they) will: רעה =feed
21 S: 9:1 WQtl	61,05	_ and feed your flocks	
		| | |<<302>>	ו =and (verbless clause)
22 S:10:1 NmCl	61,05	_ foreigners shall till your land and dress your vines	
		| |<<312>>	ו =but (front. Sub) אתם [2pM] will be: קרא =proclaim
23 S:11:1 WxYq	61,06	_ but you shall be called priests of the Lord	
		| | |<<111>>	0 (front.Sub) משמרתי אלהים

(cont.)

Ln	Sent.Cl/ Cl.type	Ch.Verse	Text of NRSV translation	Syntactic function proposed
		.	. . .	[3sM] will be: אמר =say
24	S:12:1 XYqt	61,06 . | . . . . | . |	∟ *you shall be named ministers of our God*	
			| | |	

Of importance in this procedure is the notion of experiment. We want to achieve more insight into the interaction of language system, literary composition and traditional interpretation. This may imply that one has to redo some analyses several times before one may speak of the detection of some linguistic consistency.

In an earlier version of experiments with the syntax of the book of Isaiah I worked with a text where line 14 of this chapter, which has a conjunction: *we* ('and') + *qatal* (perfect) tense form, had been connected to line 4, since that line has a *qatal* (perfect) tense form too. So the clause hierarchy of the text was a different one:

Ln	Sent.Cl.	Cl.type	Ch.Verse	Text of NRSV translation	Syntactic function proposed
				| |<<120>>	0 he has: שלח =send objSfx 2= me
4	S: 3:1	0Qtl	61,01	. |∟ *He has sent me*	
				| | |<<adjunct >><< 64>>	ל =to: חבשׁ =bind up
5	S: 3:2	InfC	61,01	. | . . | ∟ *to bind up the brokenhearted*	
				| | |<<paralDep>> <<200>> --> 64	ל =to: קרא =proclaim
6	S: 3:3	InfC	61,01	. | . . | ∟ *to proclaim liberty to the captives*	
				| | |<<ellipsis>>	ו =and (to ..)
7	S: 3:4	Ellp	61,01	. | . . | | ∟ *and release to the prisoners*	
				| | |<<paralDep>><<200>> > --> 64	ל=to: קרא =proclaim

(cont.)

Ln Sent.Cl.	Cl.type	Ch.Verse	Text of NRSV translation	Syntactic function proposed
8 S: 3:5	InfC	61,02	. | . . | . _ to proclaim the year of the Lord's favor and the day of the vengeance of our Lord | | | <<paralDep>><<200>> >> --> 64	ל =to: נחם =comfort
9 S: 3:6	InfC	61,02	. | . . | . . _ to comfort all who mourn | | | | | |<<paralDep>><<200>> >>> --> 64	ל =to: שׂים =establish
10 S: 3:7	InfC	61,03	. | . . | . . . . _ to provide for those who mourn in Zion | | | | | . . . . . |<<paralDep>><<200>> >>>> --> 64	ל =to: נתן =give
11 S: 3:8	InfC	61,03	. | . . | _ to give them a garland in stead of ashes . . . . . | | | | |<<201>>	ו =and HAVE BEEN: קרא =proclaim
14 S: 4:1	WQtl	61,03	. | . . _ They will be called oaks of righteousness, the planting of the Lord | | | | | |<<adjunct >><< 64>>	ל =to: פאר =beautify
15 S: 4:2	InfC	61,03	. | . . . . | _ to display his glory | | | |<<322>> --> 324	ו =and (they) HAVE: בנה =build
16 S: 5:1	WQtl	61,04	. | . . . . . _ They shall build up the ancient ruins | | | |	

This construction turned out to be an error. In the first place, the *qatal* tense form in line 14 is a passive form. That makes a connection with the active *qatal* form in line 4 form less likely. What is more convincing is the shift in the set of participants between line 4 ('He' and 'me') and line 14 (only 'they').

Interestingly, it was actually the second program, proposing syntactic functions, that forced me to rethink the clausal hierarchy of these lines. Due to the syntactic connection between line 14 and line 4 in the text it was reading, it proposed to interpret the lines with *qatal* in 14ff. as statements in perfect tense: 'has been' or 'have' + verb. 'It has been proclaimed to them.' 'They have rebuilt the ancient ruins.' That is very different from what existing translations propose, which translate these lines as a promise. 'They will be called oaks of righteousness, the planting of the Lord' and 'They shall build up the ancient ruins.' So there were a few options: either the program needs to learn more about Hebrew syntax or the classical interpretation in translations needs to be revised or the input data (the initial clausal hierarchy) has an error.

I tried the last option: should the *weqatal* of line 14 connect to one of the preceding infinitive clauses? Here one gets involved in the debate in classical grammars on the interpretation of the verbal form *weqatal*. Among a large number of possible interpretations one finds there the observation that indeed a *weqatal* can be used to continue an infinitive. According to P. Joüon and T. Muraoka that is the case with an infinitive with a future meaning,[14] which raises the question of how to establish a future meaning with an infinitive. In the *Hebräische Grammatik* of W. Gesenius and E. Kautzsch[15] a similar statement is made without further indication of the meaning of the infinitive. The examples presented in both grammars, however, only mention those cases where the *weqatal* immediately follows the infinitive (for example 1 Kings 2:37). Hence the experiment is undertaken with this case in the text of Isaiah 61 to try a connection between a *weqatal* clause and an infinitive clause that are located at a larger distance in the text—lines 10–14 in our text—as presented above.

This works well. The adjustment of the hierarchy, changing the clause connection from 14–4 into 14–11 resulted in a change of the syntactic interpretation. As can be seen in the presentation of the text above, the *weqatals* in line 14ff. now have an interpretation in terms of a future event or an announcement.

---

[14] P. Joüon and T. Muraoka, *A Grammar of Biblical Hebrew*, 401 (§119o 'W-Qatalti continuing an infinitive construct with future meaning').
[15] W. Gesenius and E. Kautzsch, *Hebräische Grammatik*, §§112v, 114r.

In this way the experiments with distributional and functional syntactic analysis may also contribute to the research into classical Hebrew text-level syntax.

Experiment 3. Ezekiel 36: linguistic structure—literary complexity

The text of Ezekiel 36 presents a good example of the complexity of various layers of embedded direct speech sections. The chapter is about promises of restoration of the land of Israel, its cities and its population, after the destruction by the Babylonian conquest (597–586 BC) and the deportations into exile.

There are many cases in the Hebrew Bible where readers have to question how to identify domains of direct speech in a text and to find out where a direct speech section ends,[16] so in that sense the chapter is not an exception. This situation does not create a great problem as long as one intends to parse the text in terms of clauses and clause constituents only. However, if one needs to go one step further—that is, when the goal is to find the hierarchy of clauses in the text and to establish its overall syntactic and discourse structure—one faces real analytical problems. I present the text according to a proposal for the analysis of the direct speech sections, its senders and its addressees.

Ez. 36:1   *God* → *Ezekiel*:
            'Prophesy to the mountains of Israel and say:
            Ezekiel → *mountains*:
            'Mountains of Israel, hear (plur.) the word of Yhwh:
Ez. 36:2
            Thus has the Lord said:
            God → *mountains*:
            'Since the enemy has said about you (plur.):...
Ez. 36:3   *God* → *Ezekiel*
            'Therefore, prophesy (sing.) and say:
            Ezekiel → *mountains*:
            'Since you (plur.) have been destroyed....
Ez. 36:4
            Therefore, mountains of Israel, hear (plur.) the word
            of Yhwh.

---

[16] Cf . S.A. Meier, *Speaking of speaking.*

Thus has the Lord said to the mountains...

Ez. 36:5   Therefore, thus has the Lord said:
                   God → *mountains*:
                   'I will certainly speak against the nations and Edom,
                       those who took...

Ez. 36:6   *God* → *Ezekiel*:
                   'Therefore, prophesy to the land of Israel,
                       and say to the mountains:
                   Ezekiel → *land of Israel*:
                   'Thus has the Lord has said:
                   God → *land of Israel*:
                   'In my jealousy I speak:
                   Since you had to bear ('lift up') insults,

Ez. 36:7   Therefore, thus has the Lord said:
                   I have lifted up my hand,
                   And the people will have to bear ('lift up') nsults,

Ez. 36:8   And you, mountains of Israel will bear ('lift up') your
                       fruits...'

The text repeats a number of times the instruction to the prophet to speak ('say'), which confuses the reader as to the interaction of the senders and the various addressees. The challenge is to find out how far direct speech sections are embedded into other ones, that is, how to keep track of the participants acting in this text. At what moment does the prophet, being commissioned to speak, become the sender of a direct speech section himself? Who is speaking to whom? The classical solution to overcome the complexities of the text is to assume that part of it is a later insertion that does not fit the existing discourse structure. Thus, Zimmerli in his commentary refers to Ezekiel 36:3–5 as an 'Einschub'.[17] The inserted piece of text explicitly refers to Edom, instead of only to the enemy in general, as in the other parts of the prophecy. It therefore exhibits a change in topic and in sender and addressee.

As a means to explain the history of the text this is a very acceptable proposal. However, how does one explain the text's actual structure? It presents itself as one literary text. How does one define all the domains

---

[17] W. Zimmerli, *Ezechiel*, 864–865.

of direct speech in this text? Can this text still be analysed grammatically as a coherent syntactic structure? If not, can the literary elaboration of the text still be represented in linguistic categories, such as the repeated embedding of direct speech(es)? In my proposal I have tried to do so, accepting the fact that the direct speech section in verse 2 is not completed, since verse 3 starts it again. In this way the lack of linguistic coherence remains visible in the data.

A linguistically analysed database of ancient texts has to allow for irregularities in the syntax of the texts, since one has to do justice to the history of the literary artefact. One should store the linguistic data of the text as we have it now, that is, as a text being produced not by one design, but by a tradition of readers and writers. Otherwise we have to copy the classical approach: clean up your text first before you start in order to present a proper literary unit to be analysed linguistically. But in that case you give up on your actual data and their history.

So, difficult as it may be, the challenge in e-philology is the question: can we do both—analyse and store the final literary text as a literary artefact presented to us by generations of writers and readers *and* then develop analytical devices to study its composition and the various layers to be found in it?

The distinction between linguistic analysis and literary interpretation is crucial. For example, participant tracking, based on the textual hierarchy produced by syntactic analysis will contribute to discourse analysis and at the same time will help to detect and mark cases of embedding and repetition in a text. A conclusion, however, such as this being a secondary insertion of material into the composition, based on the registration of such phenomena, remains a matter of literary interpretation. That goes beyond the actual analysis and storage of the linguistic data and should be part of additional files with annotations and comments to the textual data.

E-philology: the contrast of history and design

A classical text, especially a text composed before the print era, which has been transmitted to us by various manuscripts, can best be compared with the plan of an old city, or with a classical building: in various forms it has served many generations. Thus it has been interpreted and applied, adapted and changed, and therefore it is no longer the design of just one architect. This combination of original literary design and its history of reading and updating together have created a unique artefact that for its analysis requires a method of its own.[18] The challenge is how to integrate processes of change through history with models of human knowledge and cognition.[19] Should we understand history as if it was made by design? How do we get access to its artefacts? It is a question raised by historians[20] but until recently apparently overlooked by computer linguists. An example from my own experience illustrates that. On 14 May 1992 I participated in a symposium on computers and humanities ('Letteren en informatica in de jaren '90'). Together with J.H. Harmsen,[21] I gave a presentation of the research in Bible and computing done by the 'Werkgroep Informatica' in my Faculty of Theology. We presented some details of how we did syntactic analysis of classical Hebrew texts. The example chosen was the book of Deuteronomy. A colleague from the discipline of computer linguistics asked the question, 'How old are these texts?' I answered, 'We do not know exactly, since several generations from the seventh to the fourth century BCE have worked on the composition of this text, and many more generations since then until the Middle ages have contributed to the manuscript tradition.' But this answer clearly did not please him. 'So you are unable to do anything analytical with this corpus,' he said, 'for it has been polluted by those later hands. You have to clean up your data before any reliable analysis can be made of it.'

What he assumed was that for a proper analysis I had to rob my texts of the very history that made them so interesting as a literary corpus. Biblical texts in fact are the product of a collaboration of many scholars, not separated by location but separated in time. The ancient texts we study as artefacts from cultural and religious traditions are their common product.

When philologists interact with computer linguists, this misunderstanding frequently seems to be a problem. It is this problem which, also with the help of the colloquium that triggered this volume, we should try to turn into an exciting challenge for the methodologies of various disciplines. As soon as you enter the area of philology and

---

[18] Cf. E. Talstra, 'The Hebrew Bible and the computer'.
[19] Cf. E. Talstra, 'On scrolls and screens'.
[20] F. Ankersmit, *De sublieme historische ervaring*.
[21] See the PhD research by J.H. Harmsen, These are the Words.

ancient literature, you enter the area of history. Reading of ancient texts therefore in fact always is a kind of collaboration. It involves design, debate, and the experiences of many researchers, not located far away in place, but in time.

The task of e-philology, as I see it, is to produce text databases of ancient texts that do not need to be the outcome of our final analysis of the difficult balance of design and history, if that were even possible. Can we not rather make databases of biblical and other ancient texts that are flexible enough to allow for a study of that phenomenon—that is, the interaction of design and history?

E-philology is not restricted to a special branch of general linguistics; it should be more than the linguistic analysis of a corpus of ancient texts. Computer linguists should be willing to allow for that fact before any cooperation of IT and humanities beyond the domain of general linguistics becomes possible. Such cooperation is very necessary, especially since we want to introduce the notion and practices of experiment in our procedures of linguistic and literary analysis. E-philology is a special branch of cultural studies, because its main interest is in the interaction of design and history, that is, in the history of unique literary artefacts and the general processes of human cognition. Can computer-assisted textual analysis go with us when we enter the area of history?

References

Alkhoven, P., and P. Doorn, 'New research perspectives for the Humanities', *International journal of humanities and arts computing* 1 (2007), 35–47.

Ankersmit, F., *De sublieme historische ervaring* (Groningen, 2007).

Gesenius, W., and E. Kautzsch, *Hebräische Grammatik, Leipzig* (Leipzig, 1909[28]).

Hardmeier, Chr., E. Talstra, and B. Salzmann, *SESB: Stuttgart electronic study Bible (Stuttgarter Elektronische Studienbibel)* (Stuttgart/ Haarlem 2004; forthcoming version 3: 2009).

Harmsen, J.H., These are the words. Procedures for computer-assisted syntactical parsing and actants analysis of Biblical Hebrew texts (PhD dissertation, Amsterdam: Vrije Universiteit, 1998).

Hughes, J.J., *Bits, bytes and biblical studies. A resource guide for the use of computers in biblical and classical studies* (Grand Rapids, 1987).

Jenner, K.D., W.Th van Peursen, and E. Talstra, 'CALAP: An interdisciplinary debate between textual criticism, textual history and computer-assisted linguistic analysis', in *Corpus linguistics and textual history. A computer-assisted interdisciplinary approach to the Peshitta*, eds P.S.F. van Keulen and W.Th. van Peursen (Studia Semitica Neerlandica 48; Assen, 2006), 13–44.

Joüon, P., and T. Muraoka, *A grammar of Biblical Hebrew* (Subsidia Biblica 14/I–II; Rome, 1991).

Meier, S.A., *Speaking of speaking. Marking direct discourse in the Hebrew Bible* (Vetus Testamentum Supplements 46; Leiden, 1992).

Talstra, E., 'Desk and discipline. The impact of computers on the study of the Bible'. Opening address of the 4th AIBI Conference, in: *Proceedings of the Fourth International Colloquium Bible and Computer: Desk and discipline, in Amsterdam August 15-18, 1994* (Paris/Geneva, 1995), 25-43.

——, 'The Hebrew Bible and the computer. The poet and the engineer in dialogue', *International journal of humanities and arts computing* 1 (2007), 49-60.

——, 'On scrolls and screens. Bible reading between history and industry' in *Critical thinking and the Bible in the age of new media*, ed. Charles Ess (Lanham, Maryland 2004), 91-309.

Talstra, E., and C. Sikkel, 'Genese und Kategorienentwicklung der WIVU-Datenbank, oder: ein Versuch, dem Computer Hebräisch beizubringen', in *Ad Fontes! Quellen erfassen-lesen-deuten. Was ist Computerphilologie?* ed. Christof Hardmeier et al. (Applicatio 15; Amsterdam: VU University Press, 2000), 33-68.

Talstra, E., and C.H.J. van der Merwe, 'Analysis, retrieval and the demand for more data. Integrating the results of a formal textlinguistic and cognitive based pragmatic approach to the analysis of Deut 4:1-40', in *Bible and computer. The Stellenbosch AIBI-6 conference. Proceedings of the Association Internationale Bible et Informatique, University of Stellenbosch, 17-21 July, 2000*, ed. J.A. Cook (Brill: Leiden, 2002), 43-78.

Tov, E., *A computerized database for Septuagint studies; the parallel aligned text of the Greek en Hebrew Bible* (CATSS 2 = Journal of Northwest Semitic Languages Supplements 1; Stellenbosch, 1986).

Weil, G.E., and F. Chénique, 'Prolegomènes à l'utilisation de méthodes de statistique linguistique pour l'étude historique et philologique de la bible hébraique et de ses paraphrases', *Vetus Testamentum* 14 (1964), 341-366.

Zimmerli, W., *Ezechiel* (Biblischer Kommentar Altes Testament 13/1-2; Neukirchen, 1969).

TOWARDS AN IMPLEMENTATION OF JACOB LORHARD'S ONTOLOGY AS A DIGITAL RESOURCE FOR HISTORICAL AND CONCEPTUAL RESEARCH IN EARLY SEVENTEENTH-CENTURY THOUGHT

Ulrik Sandborg-Petersen and Peter Øhrstrøm

The word 'ontology' was constructed at the outset of the seventeenth century by Jacob Lorhard (1561–1609). In this paper, it is argued that Jacob Lorhard's work on ontology should be regarded as an important resource, not only for professional historians of science, but also for researchers working within the field of formal ontology and its use for various computational purposes. The historical background for Jacob Lorhard's ontology is outlined, and the importance of his work is discussed. In addition, we investigate some crucial features of Lordhard's ontology conceived as a systematic representation of his world view. We concentrate mainly on his religious and metaphysical views as embedded in his ontology. Finally, we discuss how Jacob Lorhard's ontology can be implemented as a digital resource for historical and conceptual research into early seventeenth-century thought.

Introduction

The study of the history of philosophical logic and scientific argumentation is valuable, not only to the professional historian, but also to the researcher working within certain branches of computer science. One of the reasons is that this field includes the study of the changes that have taken place within the systematic and conceptual approach to reality—changes which have influenced present-day research into the field of ontology. The study of ontology can be regarded as the study of the categories that exist, either in reality or in some possible world. An ontology, on the other hand, is the product of such a study.[1] Ontologies are frequently used today as an implementation technique when computational problems involving semantics have to be solved. Thus, the

---

[1] Sowa, *Knowledge representation*, 492.

study of the history of philosophical logic and scientific argumentation may help us understand both past and present problems in ontology better. For example, why were some logical or conceptual problems viewed as important in a certain historical period, whereas other problems were more or less ignored? What changes have occurred in the understanding of ontology that have led us to present-day thought on the subject?

One of the most interesting periods within the history of logic and scientific argumentation is the late sixteenth and early seventeenth century. This period lies just after the reformation and the decline of traditional scholasticism. It is also the period of the so-called scientific revolution. One of the clear characteristics of this period is its change in the focus of the study of philosophical logic and scientific argumentation. Many problems which had previously been studied with great vigour in the scholastic period were now viewed as unimportant, and therefore ignored. However, the converse was true of at least one set of conceptual problems known from traditional philosophy, which attracted a greater degree of attention in this period than it had within scholastic logic. That was the set of problems inherent in the study of ontology as the study of being. The study of being as such dates back to ancient Greek philosophy, but the term itself—ontology—was coined only much later, namely in the early seventeenth century. During this period, ontology in the sense of being was extensively studied. In addition, the problem was also redefined in a very interesting manner during this period, apparently mainly in order to support the rise of Protestant academia.

The word 'ontologia' is not found in the corpus of ancient Greek philosophical writings, meaning that it was never used in ancient philosophy. As has been suggested by Øhrstrøm, Andersen, and Schärfe,[2] the word was constructed at the beginning of the seventeenth century by Jacob Lorhard (1561–1609). Lorhard wanted, probably mainly for pedagogical reasons, to present the conceptual structure of the world—known as metaphysics—in a diagrammatical manner. Lorhard therefore used ontology as a synonym for metaphysics. However, by introducing a new term, he probably also wanted to indicate that the field was being renewed.

---

[2] Øhrstrøm et al., 'Whas has happened to ontology'.

The fact that the word 'ontology' has been imported from philosophy into modern-day computer science is fascinating in and of itself. However, even more fascinating and important is the study of the history and origins of the word, compared with present-day usage in computer science.

The understanding of ontology in computer science differs somewhat from the understanding of the word found in traditional philosophy. Whereas philosophical ontology has been concerned with the entities that exist, as well as their qualities and attributes, computational ontology has been more occupied with constructing formalized, semantic, model-theoretical, and/or logic-based models which can easily be implemented in computer systems.

In the works written within the field of philosophical ontology, Lorhard and later thinkers employed the term ontology to mean 'one singular system of thought describing being as such'—in fact intended as the one true representation of reality. Modern researchers within the field of computational, formal ontology, on the other hand, speak of sets of possible ontologies (in the plural), although they would normally regard any one single ontology as a comparably stable representation of a part of some reality. In other words, within modern formal ontology, ontologies represent multiple, or possible, fragmented but stable descriptions of some (sub)domain relative to certain selected perspectives, rather than a monolithic system. Ontologies are seen as information tools, and they do not necessarily claim any degree of truth outside the domain for which they were designed.[3] While truth is important even in the modern context, it is treated as a value that applies to selected domains or certain premises.

The questions regarding the historical development of the notions inherent in the study of being are, of course, crucial for the professional historian. But they are also important for researchers within branches of artificial intelligence and conceptual studies wanting to shed light on their work by means of a historical and broader perspective. The term ontology has become a key word within these fields of computer science. In particular, the term has become popular in relation to the Semantic Web and related technologies.[4]

---

[3] See Hesse, 'Ontologie(n)'.
[4] See Berners-Lee et al., 'The semantic web'.

In this paper, we argue that Jacob Lorhard's work on ontology should be regarded as an important resource, not only for professional historians of ideas, but also for researchers working within formal ontology and its use for various computational purposes. The rest of the paper is laid out as follows. First, we shall outline the historical background for Jacob Lorhard's ontology and the importance of his work as seen in a longer and broader perspective. Thereafter we shall discuss some crucial features of Jacob Lorhard's ontology, conceived as a systematic representation of his world view. We shall mainly concentrate on his religious and metaphysical framework embedded within his ontology. In the final section we shall discuss how Jacob Lorhard's ontology can be implemented as a digital resource for historical and conceptual research into seventeenth century thought. This digital resource will open up Lorhard's work to professional historians and researchers within the field of formal ontology who want to see their work in a broader perspective, and thereby encourage transdisciplinary debate on the nature of ontology.

Jacob Lorhard's ontology in historical perspective

Jacob Lorhard was born in 1561 in Münsingen in Southern Germany. We know little of his life, but it does appear that he met Johannes Kepler at Tübingen University, where Kepler is known to have studied in the period 1587–1591. Lorhard was deeply interested in metaphysics, conceived as the study of the conceptual structure of the world. In 1597, he published his Liber de adaptione, in which he wrote:

> Metaphysica, quae res omnes communiter considerat, quatenus sunt ὄντα, quatenus summa genera & principia, nullis sensibilibus hypothesibus subnixa.[5]

> Metaphysica, which considers all things in general, as far as they are existing and as far as they are of the highest genera and principles without being supported by hypotheses based on the senses. (Our translation.)

Lorhard came to the Protestant city of St. Gallen in 1602, where he worked as a teacher and preacher. The following year, in 1603, he became 'Rektor des Gymnasiums' in St. Gallen. He was accused of

---

[5] *Liber de adaptione*, 75.

alchemy and also of having a heretical view on baptism. He was, however, able to defend himself convincingly against his accusers, and his statements of belief were in general accepted by the church in St. Gallen.[6]

In 1606, Lorhard published his *Ogdoas Scholastica*, a volume consisting of eight books dealing with the subjects of Latin and Greek grammar, logic, rhetoric, astronomy, ethics, physics, and metaphysics (or ontology), respectively see Plate 8). The book was apparently intended as a school book, but it should also be seen as an attempt at presenting the whole corpus of academic learning in a clear manner.

Although Lorhard only used his new term ontology a few times in his book, he did present his new term in a very prominent manner, letting 'ontologia' appear on the frontispiece of *Ogdoas Scholastica*. As has been argued in by Øhrstrøm, Andersen, and Schärfe,[7] this was probably the first published use of the term ontology ever. The phrase 'Metaphysices seu ontologiæ' in the title of the book indicates that 'ontologia' is to be used synonymously with 'metaphysica'.

After Lorhard's death in 1609 at least some of his ideas were taken up by Rudolf Göckel (1613) and introduced to Protestant academia in Germany. As convincingly argued by Claus Asbjørn Andersen, Göckel's presentation of ontology includes an even stronger emphasis on the importance of abstraction from the material.[8] In this way ontology may be characterized as the study of what can be understood by the human intellect, but organized in a system that reflects the ordering of conceptual understanding in a proper manner. It is an important guiding principle in Göckel's ontology that the fundamental terms of that structure are organized as concept pairs. The same is clearly the case in Lorhard's ontology. His system is presented in terms of dichotomies whenever possible: he clearly sought to divide each complex concept into two subclasses characterized by contradicting terms (see Plate 9).

Øhrstrøm et al. have argued that Lorhard's ontology was very much inspired by Clemens Timpler of Heidelberg, whose *Metaphysicae systema methodium* was published in Steinfurt in 1604 and in Hanau in 1606.[9] Timpler 'proposed that the subject-matter of metaphysics is not

---

[6] See Hoffmeier et al., *Alchemie in St. Gallen*, 28 ff. and Bätscher, *Kirchen- und Schulgeschichte*, 171 ff.

[7] Øhrstrøm et al., 'What has happened to ontology'.

[8] Andersen, *Philosophia*: 96 ff.

[9] Øhrstrøm et al., 'Jacob Lorhard's ontology'.

being, but rather the intelligible, παν νοητον'.[10] He claims: 'metaphysica est ars contemplatiua, quae tractact de omni intelligibili, quatenus
ab homine naturali rationis lumine sine ullo materiae conceptu est
intelligibile' (metaphysics is a contemplative art which treats of every
intelligible thing, to the extent that it is intelligible by men through the
natural light of reason without any concept of matter).[11] Timpler's work
was enormously influential on Lorhard's ontology. Timpler's *Metaphysicae* is however written in a more traditional style than Lorhard's
textbook: it is divided into five books, of which each chapter presents
an aspect or a part of his metaphysical views, followed by a number
of questions and answers dealing with the philosophical issues arising
from the distinctions offered at the beginning of each chapter. For
example, q. 5 of bk. 1, cap. 1 is 'what is the proper and adequate subject matter of metaphysics?,' to which the answer, naturally, is 'omne
intelligibile', or 'every intelligible thing'. The divisions and distinctions
in Timpler's work can be found almost universally without change in
Lorhard's ontology, with the exception that Lorhard's text omits all
the philosophical comments and considerations.[12] What is interesting
is that in many places where Timpler raises questions about his classification and characterization, Lorhard adopts his distinctions without
indicating that they might be questionable. However, by his extensive
use of digrams Lorhard presents the material in a much more elegant
and readable manner than Timpler.

Following Timpler's definition of metaphysics, Lorhard defined
ontology as 'the knowledge of the intelligible by which it is intelligible'.[13]
His ontology is hence a description of the world of intelligibles, i.e.,
the items, concepts, or objects which are understandable or conceivable from a human perspective. The emphasis on the intelligibility
of the world is essential in Timpler's and Lorhard's ontology. When
Lorhard followed Timpler's lead and adopted this new proposal about
the subject matter of metaphysics, or ontology, he agreed with the idea
that we in formulating ontology are concentrating on the knowledge
by means of which we can conceive or understand the world. In this
way ontology is seen as a description of the very foundation of scientific activity as such. Lorhard holds that the human rationality must

---

[10] Timpler, *Metaphysicae*, 635.
[11] Bk. 1, cap. 1.
[12] For a comparative taxonomy, see Lamanna, 'Correspondences'.
[13] Uckelman, Diagraph, 3.

function on the basis of what he and Timpler both call 'the natural light of reason'.[14] Ontology captures this fundamental understanding of the basic features of the world. Based on this knowledge everything else—to the extent that it is intelligible at all—becomes conceivable. This approach presupposes that there is in fact only one true ontology—the one that reflects the world as it truly is. This belief was in fact very important for the rise of modern science in the early seventeenth century. According to J. Needham, the confidence that an order or code of nature can in fact be read and understood by human beings was one of the important cornerstones for the rise of modern science in Europe.[15] This strong belief was absent in Eastern civilizations in the early seventeenth century. Lorhard, again following Timpler, divides the world of intelligibles into two parts: the universals and the particulars. The set of universals can be further divided into two parts: the set of basic objects, and the set of attributes.[16]

Øhrstrøm et al. 2007 have argued that Lorhard's approach to ontology and in particular his use of diagrams was probably very much inspired by Peter Ramus (1515–1572), who had argued that for pedagogical reasons scientific knowledge should be simplified using diagrams, and preferably organised in dichotomies.[17] Certainly it is obvious that Lorhard's use of diagrams in his book is very close the way Peter Ramus used diagrams in his works. Like Ramus, Lorhard accepted the idea that we may understand reality (or at least important aspects of reality) by means of the natural light of reason, i.e., we have as rational beings access to necessary truth in mathematics and in the world in general.

In 1562 Ramus converted to Calvinism, and he was murdered in Paris in the St. Bartholomew's Massacre on August 26, 1572. The fact that he was considered to be a Protestant martyr made many intellectual Protestants interested in his ideas. In fact, his religious and scientific ideas became very influential in the Protestant world during the sixteenth and seventeenth century. As emphasized by R. Hooykaas Ramus was interested in how the making of instruments could support the application of mathematics in the study of reality.[18] This

---

[14] Uckelman, Diagraph, 3
[15] Needham, *The grand titration.*
[16] Uckelman, Diagraph, 3.
[17] Øhrstrøm et al., 'Historical and conceptual foundation'.
[18] Hooykaas, 'The rise of modern science'.

interest was probably based on the belief in a mathematical structure of the physical and conceptual universe. When taken together with the practical approach to mathematics, this view turned out to be essential for the rise of modern natural science. Timpler worked in the same tradition, and Lorhard followed in the footsteps of both Timpler and Ramus by making the research programme more clear and proposing the term 'ontology' as new label for the renewed approach to the study of being as such.

Some crucial features of Lorhard's world view as represented in his ontology

Jacob Lorhard's ontology is a very interesting work. Unfortunately, however, it is little known. For several reasons it should be made available to a broader audience of contemporary academics. First of all, Lorhard's book is clearly relevant as the first book in which the word 'ontology' occurs. Coining this term and illustrating how it should be used may be seen as an attempt at contributing to the establishment of a Protestant academia. Secondly, Lorhard's ontology is a very useful and precise resource for anyone who wants to study the Protestant world view in the early seventeenth century. Thirdly, Lorhard's ontology is a nice illustration of the use of the systematic teaching techniques composed in the diagrammatical style of Peter Ramus.

Let us consider some examples from Lorhard's book on ontology in order to illustrate his strong belief in a strict or logical order of everything. We shall concentrate on examples dealing with a metaphysical or even religious approach to reality. It is, however, interesting to notice the way the word 'rational' is used by Lorhard. According to him the order of beings may be either real or rational.[19] If it is real, it does not depend on human cognition, but exists independently of human observation and cognition. A purely rational order, on the other hand, is mind-dependent: it exists as a result of human reason. Lorhard's ontology contains a rather elaborate description of the various kinds of order of being (see Figure 1).

---

[19] Uckelman, Diagraph, 41.

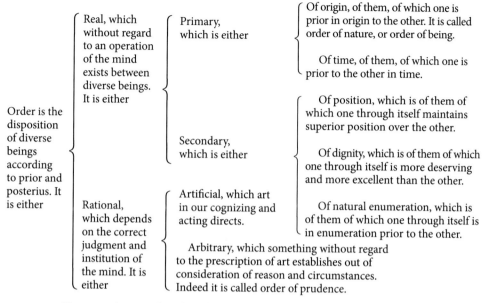

Figure 1: A part of Lorhard's ontology, dealing with order of being.[20]

Lorhard clearly believed in an essential 'order of nature' related to the origin or creation of the world. This order is structural. It is, however, important to emphasize that the order should not be understood as something static or inescapable. As we shall see, the order should be understood in the perspective of time.

Lorhard also deals with morality in his ontology. He claims that in some cases, moral qualities are just apparent or imaginary. In other cases, however, a being or an intelligible thing is in fact 'good in reality'.[21] It should be noted here, that the alternative to 'imaginary' is not 'real' as it was in the distinction between 'rational' and 'real' mentioned above. When it comes to ethical categories, Lorhard instead employs the term 'true' (Latin: *verus*) when dealing with mind-independent qualities, claiming that mind-independent goodness is, in fact, true in itself. Similarly, malice is described in Lorhard's ontology as being either true or apparent. If a malice is true, then it is truly bad

---

[20] After Uckelman, Diagraph, 41.
[21] Uckelman, Diagraph, 17.

Goodness
is an actuality
of good,
because it is
good; or it
is a quality,
through
which a
Being is
denominated
good. It is
either

Apparent, which is an imaginary quality, through which a Being is seen to be good, but in truth is not good.

True, which is a real quality, through which a Being is good in reality, when it is seen [to be good]. It is either

Absolute, through which a Being in truth is good in itself without respect to anything. It is either

Respective.

Infinite, or primary which is in the sole uncreated Being. It is the source of every finite good.

Finite or secondary, which is in a created Being, [which is good] as much as the same things are impressed in the image of the first good through participation. It is either

Natural, which is the agreement in a being with the rule of creation, or of a generating Nature.

Moral, which is an agreement of a Being with moral law.

Artificial, which is an agreement of a Being with the rule of art.

Figure 2: A part of Lorhard's ontology, dealing with morality.

in itself and without respect to anything else. If it is mind-dependent, however, then the malice is merely apparent (see Figure 2).

It is worth noting that according to Lorhard there is in fact something which is good in itself and also something which is bad in itself. This means that he accepts the idea of an absolute ethics, which is not a construction of human rationality, but which on the other hand can be understood or realized by humans.

Lorhard's framework is basically Aristotelian. Following this tradition, Lorhard claimed that the understanding of causality is essential for the understanding of the world. This point was formulated by Aristotle himself in the following manner:

> Knowledge is the object of our inquiry, and men do not think they know a thing till they have grasped the 'why' of it (which is to grasp its primary cause).[24]

---

[22] Uckelman, Diagraph, 17.
[23] After Uckelman, Diagraph, 17.
[24] Aristotle, Physics II.3.

In order to establish a theoretical framework corresponding to the structure of the world including its temporal relations, he refers to the four Aristotelian causes:[25]

> Cause by reason of efficiency,
> Cause by reason of matter,
> Cause by reason of form,
> Cause by reason of finality (telos).

For some reason Lorhard has chosen to list these four causes in an order different from the one used by Aristotle himself.[26]

From a modern perspective, the most controversial part of his treatment of causality is probably that he views final causes (telos) as existing not only by means of human intent. On the contrary, according to Lorhard's world view, teleological causes may also be found in nature. There can be little doubt that this approach to purpose (telos) in nature should be interpreted in light of the religious assumptions incorporated in Lorhard's ontology. According to Lorhard, teleological causes may also exist in nature by means of divine intent.

The temporal aspect of Lorhard's ontology can be found not only in his emphasis upon the importance of causality. It is also evident from the fact that the essence of a being is introduced in terms of positive characteristics. According to Lorhard, a being is an actuality or performance (Latin: *actus*).[27] If a being is real, the actuality or performance constituting its essence will reflect features in the external world taking place independently of human cognition. If the being is not real, the actuality or performance in question will depend on human cognition. In both cases the reference to 'actus' must involve some kind of process. This clearly means that there is an aspect of temporality involved in the essence of beings.

In Lorhard's world view, actions of God may be divided into two kinds, namely internal and external. The internal actions of God are those which take place inside of God himself, such as his thinking, willing, and loving. The external actions of God are those which have one or more effects in objects outside of God. The external actions of God are further subdivided into general and special actions. The special

---

[25] Uckelman, Diagraph, 25.
[26] Aristotle, *Physics* II.3.
[27] Uckelman, Diagraph, 4.

actions are those in which he deals with understanding creatures, such as man. Lorhard assigns many of the concepts from systematic theology under these external, special actions, including redemption, regeneration, justification, and salvation.

According to Lorhard, the external, general actions of God can be subdivided into either eternal or temporal actions. The temporal ctions are further subdivided into the first cause, namely God's creation of the world, and its management. His management of the world is further subdivided into the ordinary and the extraordinary actions. The ordinary actions are those by means of which he commonly preserves the order of the universe. The extraordinary actions must be seen as miraculous actions, but according to Lorhard, the purpose of the extraordinary actions is the same as that of the ordinary actions, namely to administer and conserve certain things of the world. The implication is that the order of the universe is preserved by God in external, general, temporal actions, and miraculous actions only differ from ordinary actions in their lack of commonness. The idea that God preserves the order of the universe may be found in the Bible in Hebrews 1:3, and is reflected here in Lorhard's ontology. As mentioned previously, Lorhard was known as a preacher and theologian, which is clearly reflected in his analysis of God's actions.

Lorhard divides God's eternal actions into preordination, as the determination of the general plan for the temporal world, and precognition, as the act by which God knows anything that is going to happen in the future whether it is necessary or contingent. This precognition may be related to Lorhard's characterization of God's supreme perfection, one kind of which is supreme 'wisdom or omniscience…according to which he is perfectly wise, subject to no error of ignorance.'[28]

It should be noted that all the external divine actions mentioned above are carried out from a standpoint outside the temporal world. The idea seems to be that there is a non-temporal dimension of existence, logically prior to time, from which God can relate to the temporal world. From this non-temporal or eternal dimension, God can either act in eternity by planning the course of events (preordination) or by seeing it (precognition).

---

[28] Uckelman, Diagraph, 45.

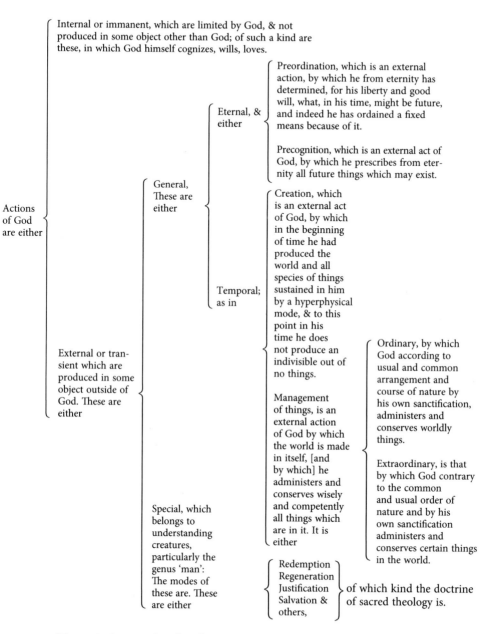

Figure 3: A part of Lorhard's ontology dealing with actions of God.[29]

---

[29] After Uckelman, Diagraph, 46.

The discussion of the logical tension between preordination and precognition was very important during the reformation and the counter reformation. The secular version of this problem is in fact still considered to be one of the most important problems within temporal logic.[30] Lorhard's very brief diagrammatical statement about the problem is that both preordination and precognition should be understood as eternal actions of God, as opposed to temporal actions.

Implementing Jacob Lorhard's ontology as a digital resource

As has been argued by Øhrstrøm, Andersen, and Schärfe,[31] the import of the word 'ontology' from philosophy into computer science is not a case of immediate transfer, and as such is not easy to describe briefly. Many sources of influence and many trends in knowledge representation have contributed to this import.[32] The number of conferences, papers, and books on the subject of ontology within the (sub)fields of computer science makes it difficult to talk about ontology in computer science as an easily definable concept: the use of the word 'ontology' is polysemous to a high degree. For an excellent overview, see Roberto Poli's 'Framing Ontology'.[33]

Presumably, it was John McCarthy who first introduced the term ontology in the literature of the subfield of computer science called Artifical Intelligence. In 1980, McCarthy used the term in his paper on circumscription in a discussion of what kinds of information should be included in our understanding of the world. McCarthy introduced ontology as 'the things that exist'.[34] His use of the term may be seen as a widening of the concept, but in a non-invasive manner. In contrast, one of the most frequently cited definitions of ontology is that of 'a specification of a conceptualization', found in Gruber's paper from 1993, which very much changes the view of ontology and ontology research from that held by the field of philosophy.[35] This definition has often been debated,[36] but clearly, it alludes to something much more

---

[30] Øhrstrøm, 'In defence of the thin red line'.
[31] Øhrstrøm et al., 'What has happened to ontology', 434.
[32] See Orbst and Liu, 'Knowledge representation', and Welty, 'Ontological research'.
[33] Poli, 'Framing ontology'.
[34] McCarthy, 'Circumscription', 31.
[35] Gruber, 'A translation approach'.
[36] See for instance Guarino, 'Formal ontology'.

subjective and changeable than the classical view allows for. In 1993, when Gruber published this paper, the term had already gained some popularity, and at that time several sources were available.[37] In 2000, John Sowa suggested the following definition:

> The subject of ontology is the study of the categories of things that exist or may exist in some domain. The product of the study, called 'an ontology', is a catalog of the types of things that are assumed to exist in a domain of interest, D, from the perspective of a person who uses language L for the purpose of talking about D.[38]

Sowa's definition implies that, viewed within knowledge engineering, a theory of ontology has to be seen in relation to a certain domain of interest and it also must presuppose a certain perspective. Lorhard's ontology, on the other hand, is intended as a possibly exhaustive catalogue of the categories that exist in the world. His perspective is that of a Protestant academic from the seventeenth century.

Lorhard's ontology was translated into English by Sara L. Uckelman.[39] In addition, the English text has been transformed into two different hypertext versions. The problems related to translating the ontology from Renaissance Latin to English is discussed in the annotated translation. Here, we shall report on some of the most interesting aspects of producing the hypertexts, some of which relate to the meaning of the brackets.

In modern ontologies, as in Lorhard's ontology, distinctions are made based on relations that are said to inhere between the concepts or types in the ontology. These relations are sometimes ill defined. For example, in a given ontology, it may be unclear whether such a relation is an 'is-a' relation, a 'has-a' relation, an 'is-part-of' relation, a 'has-role' relation, or a 'has-attribute' relation, to name but a few possible relations. If the kind of relation is not formally specified, it may be difficult to decipher the intent of the author of a given ontology in their distinctions. Deciphering the intent of the author of an ontology is obviously very important when interpreting such an ontology, and here the nature of the distinctions—the relations that inhere between the concepts in the ontology—is of utmost importance.

---

[37] Welty, 'Ontological research'.
[38] Sowa, *Conceptual structures*, 492.
[39] Uckelman, Diagraph.

In the case of Lorhard's ontology, it is our opinion that the Ramean brackets employed by Lorhard mean different things at different points in the ontology. In general, we hold that the brackets (when splitting a concept in two or more, i.e., '{') denote the relation 'is-a', i.e, the relation is a subtype relation. However, when one of the subordinate texts is introduced by the Greek word *logos*, it is our opinion that the brackets do not mean 'subtype', but rather that the logos text is an explanation or commentary on the other subordinate text.

The use of right brackets ('}') is very widespread throughout Lorhard's ontology, and indeed throughout the *Ogdoas*, and an example occurs in Figure 3 above.[40] The left brackets always seem to imply a disjunctive relation, whereas the right brackets always seem to imply a conjunctive relation. However, the semantics of this notation was not initially clear to us, and we did in fact speculate as to whether the right brackets could be seen as a forerunner of multiple inheritance. A closer study of the original texts does, nonetheless, reveal that the right bracket is a shorthand notation for a simple tree structure.

A contemporary version of Lorhard's text could be a simple hypertext as suggested by Sara Uckelman.[41] This implementation remains true to the original and preserves the structure in a very direct manner. Possible non-invasive additions could include more navigational aids such as a bi-directional link structure to help maintain the awareness of the big picture. It would also be desirable to have the original book pages shown alongside the translation. In addition, it would be advantageous to have the text available both in the English translation and in the original Latin and Greek, with easy navigation between the two texts.

Our own implementation[42] of the English translation as a hypertext is based on the LaTeX source of the English translation made by Sara Uckelman. The process of generating the hypertext is fully automatic, and proceeds as follows. First, the LaTeX source is translated into XML using a pipeline of programs written in the scripting languages 'sed' and 'Python'. This pipeline applies a series of transformations to the LaTeX source, resulting in an XML version of the text that has been enriched with hyperlinks where applicable. Finally, the

---

[40] From Uckelman, Diagraph, 46.
[41] In Uckelman, Diagraph.
[42] See http://www.hum.aau.dk/~poe/Lorhard/id00001.html.

XML version is read by another program which builds an in-memory structure representing the ontology as a directed acyclic graph, the nodes of which are then 'dumped' as individual HTML pages. Each HTML page has the following information: the text of the entry itself; the text of its parent (if any; the root of the so-called diagraph does not have a parent, of course), combined with a link to the parent HTML page; and finally, link(s) to the sub-entry or sub-entries of the entry being viewed.

Given that we have an in-memory representation of the ontology as a diagraph, it would be interesting to add some of the following features to the implementation. For example, it would be advantageous to be able to navigate the ontology, not as individual pages, but as one graph on a single page, much like a view of a street map that scrolls when the user drags the mouse. That way, the text and its relations would be both searchable and easier to navigate. Another advantageous enhancement might be, as suggested above, to have the Latin and Greek text within easy reach, for example, as pop-ups. A third advantageous enhancement might be to store the hypertext in an information retrieval system, such as Emdros,[43] for easy searching based on occurrences of words. Given Emdros's capabilities for structural search, it would also be easy to implement the diagraph in such a way that parent-relationships could be searched. For instance, it might be interesting to know which concepts included the word 'actuality' while also being a descendant of a node which included the term 'general'. Such searches could be made if the diagraph were represented as an Emdros database. The most pressing need of enhancement to our current implementation is to make the hypertext easier to use and navigate. If enhancing the usability of our hypertext is to be done correctly we will have to involve other users, which makes it a considerable challenge.

Conclusion

An implementation of Jacob Lorhard's ontology can become an important digital resource by means of which digital creativity may be facilitated within the study of early seventeenth-century thought,

---

[43] Petersen, 'Emdros', and 'Principles, implementation strategies'.

and an inspirational background to research in modern formal ontology. The historian interested in early seventeenth-century thought may want to look into the details of Lorhard's ontological commitments, and the computer scientist working with formal ontology may want to compare Lorhard's ontological commitments with the modern understanding of everything. In both cases the implementation of Jacob Lorhard's ontology will probably be an important digital tool from which researchers may benefit, provided that a better user-experience is designed and implemented, and provided that the original Latin and Greek text and its English translation are integrated in an accessible manner.

References

Aristotle, *Physics*, Books 1–4 (Tra. P.H. Wicksteed and F.M. Cornford, Loeb Classical Library, 1929).

Andersen, Claus Asbjørn, *Philosophia de ente seu transcendentibus. Die Wissenschaft vom Seienden als solchem und von den Transzendentalien in der spätscholastischen Metaphysik. Eine Untersuchung im Ausgang von Franciscus Suárez* (Konferens-speciale, Københavns Universitet, 2004).

Berners-Lee, T., J. Hendler, and O. Lassila, 'The semantic web', *Scientific American* 284, 5 (May 2001), 34–43.

Bätscher, Theodor Wilhelm, *Kirchen- und Schulgeschichte der Stadt St. Gallen*, Erster Band 1550–1630 (St. Gallen: Tschudy-Verlag, 1964).

Göckel (Goclenius), Rudolf, *Lexicon philosophicum, quo tanquam clave philosophicae fores aperiuntur* (Francofurti, 1613; reprographic reproduction, Hildesheim: Georg Olms Verlag, 1964).

Gruber, T.R., 'A translation approach to portable ontologies', *Knowledge acquisition* 5, 2 (1993), 199–220.

Guarino, N., 'Formal ontology in information systems', in *Proceedings of the first international conference (FOIS'98), June 6–8, Trento, Italy*, ed. N. Guarino (Amsterdam: IOS Press, 1998), 3–15.

Hesse, Wolfgang, 'Ontologie(n)', GI Gesellschaft für Informatik e.V.—Informatik-Lexikon (2002), available online from the GI-eV website: http://www.gi-ev.de/informatik/lexikon/inf-lex-ontologien.shtml.

Hoffmeier, T., U.L. Gantenbein, R. Gamper, E. Ziegler, and M. Bachmann, *Alchemie in St. Gallen* (St. Gallen: Sabon-Verlag, 1999).

Hooykaas, R., 'The rise of modern science. When and why?', *British journal for the history of science* (1987), 453–73.

Lamanna, M., 'Correspondences between the works of Lorhard and Timpler', Bari University (2006), http://www.formalontology.it/essays/correspondences_timpler-lorhard.pdf

Lorhard, Jacob. *Liber de adeptione veri necessarii, seu apodictici* (Tubingae, 1597).

——, *Ogdoas scholastica* (Sangalli, 1606).

McCarthy, J., 'Circumscription. A form of nonmonotonic reasoning', *Artificial intelligence* 13, (1980), 27–39.

Needham, J., *The grand titration. Science and society in East and West* (London: Allen & Unwin, 1970).

Orbst, L. and H. Liu, 'Knowledge representation, ontological engineering, and topic maps', in *XML Topic Maps. Creating and using topic maps for the Web*, ed. J. Park (Boston: Addison Wesley, 2003), 103–148.

Petersen, U., 'Emdros. A text database engine for analyzed or annotated text', ICCL, Proceedings of COLING 2004, held August 23–27 in Geneva (International Commitee on Computational Linguistics, 2004), 1190–1193, http://ulrikp.org/pdf/petersen-emdros-COLING-2004.pdf.

——, 'Principles, implementation strategies, and evaluation of a corpus query system', in: Finite-state methods and natural language processing 5th International Workshop, FSMNLP 2005, Helsinki, Finland, September 1–2, 2005, Revised papers, eds Anssi Yli-Jyrä, Lauri Karttunen, and Juhani Karhumäki (LNCS 4002; Heidelberg and New York: Springer, 2006), 215–226, http://www.hum.aau.dk/~ulrikp/pdf/Petersen-FSMNLP2005.pdf.

Poli, R., 'Framing ontology' (2004), available online from the Ontology resource guide for philosphers: http://www.formalontology.it/essays/Framing.pdf.

Schärfe, H., 'Narrative ontologies', in *Knowledge economy meets science and technology—KEST2004*, eds C.-G. Cao and Y.-F. Sui (Beijing: Tsinghua University Press, 2004), 19–26.

Sowa, John F., *Conceptual structures. Information processing in mind and machine* (Reading, Mass.: Addison–Wesley, 1984).

——, *Knowledge representation. logical, philosophical, and computational foundations* (Pacific Grove, CA: Brooks Cole Publishing Co., 2000).

Timpler, C., *Metaphysicae systema methodicum* (Steinfurt 1604).

Uckelman, S.L. (tra.), Diagraph of metaphysics or ontology by Jacob Lorhard, 1606 (2008), http://staff.science.uva.nl/~suckelma/lorhard/lorhard-english.html.

Welty, C., 'Ontology research', *AI magazine* 24, 3, (2003), 11–12.

Øhrstrøm, P., J. Andersen, and H. Schärfe, 'What has happened to ontology', in *Conceptual structures. Common semantics for sharing knowledge. 13th International Conference on Conceptual Structures, ICCS 2005, Kassel, Germany, July 17–22, 2005. Proceedings*, eds F. Dau, M.-L. Mugnier and G. Stumme (LNCS 3596; Berlin and Heidelberg: Springer-Verlag, 2005), 425–438.

Øhrstrøm, P., S.L. Uckelman, and H. Schärfe, 'Historical and conceptual foundation of diagrammatical ontology', in *Conceptual structures. Knowledge architecture for smart applications. 15th International Conference on Conceptual Structures, ICCS 2007, Sheffield, UK, July 22–27, 2007. Proceedings*, eds U. Priss, S. Polovina, and R. Hill (LNCS (LNAI) 4604, Heidelberg and New York: Springer, 2007), 374–386.

Øhrstrøm, P., H. Schärfe, and S.L. Uckelman, 'Jacob Lorhard's ontology. A seventeenth century hypertext on the reality and temporality of the world of intelligibles', in *Conceptual structures. Knowledge visualization and reasoning: 16th International Conference on Conceptual Structures, ICCS 2008, Toulouse, France, July 7–11, 2008. Proceedings*, eds P. Eklund and O. Haemmerlé (Lecture Notes in Computer Science 5113, Berlin, 2008), 74–87.

Øhrstrøm, P., 'In defence of the thin red line. A case for Ockhamism', *Humana.Mente* 8 (2009), 17–32.

PART TWO

SCHOLARLY AND SCIENTIFIC RESEARCH

CRITICAL EDITING AND CRITICAL DIGITISATION

Mats Dahlström

This chapter discusses scholarly editing based on textual criticism and its transmission mechanisms, science ideals, and socio-cultural functions. Primarily, its pattern of conflicts is mapped—historically and presently. It is suggested that this pattern is similar to that of other transmission activities within socio-cultural institutions. The chapter then proceeds to deal with library digitisation as a transmission activity. Two digitisation approaches are discussed: the currently much debated mass digitisation, and an approach that bears many similarities to textual criticism and that this chapter therefore suggests we refer to as critical *digitisation. The relation between these two digitisation strategies is analysed. Finally, the chapter returns to scholarly editing and analyses its relation to library digitisation, with a particular focus on comparing the similarities and differences between these two transmission activities.*

Scholarly editing and transmission ideals

Scholarly editing based on textual criticism is a bibliographical, referential activity.[1] It examines a bulk of documents, compares their texts, normally clusters these around the abstract notion of a work, arranges them in a web of relations, and attempts to embody this web in the scholarly edition. The edition, then, becomes a surrogate purporting to represent, tag and comment upon the edited work. In a sense, the editor reproduces existing documents by making a new document that also embodies a documentation of the textual history and the editorial process.

At the same time, scholarly editions are hermeneutical documents and subjective interpretations, in two senses: they carry with them an ideological and hermeneutical heritage, and they also exert an interpretative influence over the objects they are designed to embody and represent. Nevertheless editions by tradition pretend to convey a sense of value-free objectivity, a mere recording of facts.

---

[1] Parts of this chapter are based on a previous anthology chapter in Swedish: M. Dahlström 'Kritisk digitalisering', in *Fra samling til Sampling: Digital formidling of kulturarv* eds N. D. Lund et al. (Copenhagen: Multivers, 2009), 171–191.

This difference between thinking of scholarly editing as either subjective interpretation or as the objective reporting of scientific facts is one of many interesting tensions and potential conflicts within scholarly editing. In fact, when studying the history of editing and textual criticism, one can easily come up with a long list of such tensions and conflicts. I list a few of them in Table 1, and discuss them briefly in the text that follows.

Table 1: Tensions in scholarly editing

Tensions in scholarly editing	
critical	non-critical
interventionist	non-interventionist
interpretative	factual
facts *as* interpretation	facts separable from interpretation
ambiguous	disambiguable
idiographic	nomothetic
contingent tools	universal tools
material document	abstract text
the one text: discriminatory	the many texts: comprehensive

A long established distinction, firstly, is the one between scholarly editing that is *critical* and scholarly editing that is *non-critical*—the latter exemplified by documentary editing or facsimile versioning.[2] At times non-critical editing is even looked upon as more or less mechanical and trivial transmission. One might however argue that digital scholarly archives displaying full-text versions in parallel, with no single established text in the centre, threaten to break down this distinction. In creating such a digital archive, the editors open up the doors to their editorial 'lab', so to speak, and turn it into an archive that might be cumulative and be run jointly with other editors and scholars and allow them to access source document representations for new editorial endeavours.[3] A much discussed feature in such archives is the idea of abandoning the established critical text as a privileged gateway to the complex of versions, and handing over this task to the individual

---

[2] See for example D. Greetham, 'Textual scholarship'. See also A. Renear, 'Literal transcription'.
[3] G. Bodard and J. Garcés, 'Open source critical editions'.

user. What is really left of the concept of the scholarly edition, if the critically established base text is removed?

As will be argued later in this chapter, the boundary is further blurred between critical editions and some of the digital facsimiles that libraries produce and that might arguably be considered as 'critical'. Image management (such as capture and subsequent editing) is perhaps the digital editing phase where the presumed distance between objectivity and subjectivity is at its largest. On the one hand, image transmission in general and facsimile production in particular has traditionally been regarded by textual critics as non-critical activities, where the editor supposedly recedes into the background, and where the user is brought closer to the source documents by having 'direct' access, as it were, to the originals. On the other hand, digitisation and the subsequent editing of images has perhaps more than any other editing phase made us attentive to the fact that virtually all parameters in the process (image size, colour, granularity, bleed-through, contrast, layers, resolution etc.) require intellectual, critical choices, interpretation, and manipulation.

A related distinction is therefore that between the acknowledged presence and the presumed absence of the editor. This is the fundamental issue of *intervening* or *not intervening* in the text, something Greetham referred to as the Alexandrian and the Pergamanian editorial ideals.[4] The former accepts and even presupposes intervention and corrections, laying the ground for eclectic editing. To the latter, however, interventions and corrections are theoretically awkward (and even come close to heresy), making way for the school of facsimile and best-text editing. If in academic discourse scholars appear to 'hide' their role as narrative writers,[5] then digital archives based entirely on diplomatic and facsimile editions promise the disappearance of editors altogether, inviting readers to step in and fill the creative, authoritative editorial function.

Then there are as well tensions between different scholarly and scientific *ideals*. The edition can on the one hand be thought of as a

---

[4] D. Greetham, *Theories of the text.*

[5] Charles Bazerman (*Shaping written knowledge*) makes the following observation on the scientific article as written genre: '[T]o write science is commonly thought not to write at all, just simply to record the natural facts. Even widely published scientists, responsible for the production of many texts over many years, often do not see themselves as accomplished writers, nor do they recognize any self-conscious control of their texts.' This seems to be a legacy in the scholarly editing community as well.

scientific, value-neutral and objective report, the 'calculus' if you will. On the other hand, there are powerful arguments in favour of viewing the edition as a scholarly, interpretative, subjective statement. In the one case, the edition's text is primarily recognised as a scientific fact, and in the other as an interpretation. Similarly, there is a tension between the notion that facts *are* interpretations and the idea that we can *separate* facts from interpretation, such as in XML encoding (separating between accidental form and substantive content), in stand-off markup, in other means of producing descriptions and separating them from the objects they describe, as well as in synoptic full versions of facsimiles that represent no tampering and provide 'raw' material. So there is a split between thinking that the objects (and tools) of editing *can* or *cannot* be subjected to universally agreed disambiguation.[6] And if the content can be decontextualised and disambiguated, it can also cumulatively form building blocks in other and different types of editorial endeavours with little or no problem.[7] In consequence, editions are in this respect either stores of scholarly raw material that support future reusability by other editors and scholars, or argumentative and context-bound statements.

Another tension between scholarly ideals is the one between viewing editing as a primarily *nomothetical* or *idiographical* affair. The former maps patterns of common, regular and predictable traits in large amounts of text, while the latter rather wants to highlight the unique, the different, the contingent. This creates a subordinated tension between attempting to design *universal, project-general* or *contingent, project-specific* tools.[8] It is also somewhat related to another, much recognised distinction in scholarly editing, namely between different conceptions of the very empirical object of editing: either the text as reduced to linguistic sign sequences or as a meaning conveyed by those linguistic sequences in conjunction with layout, typeface, colour and the rest of the graphical and material appearance that the document

---

[6] For instance, the notion of digital do-it-yourself-editions implies that the digital documents produced (the 'target' documents) are in all respects equivalent to the source (or 'departure') documents, and that the user, granted with direct access to the same source documents as the editor had, can tread different paths and make different choices from those of the editor. It also assumes that the transmission process has been able to be disambiguated, leaving the target documents therefore unaffected by any significant distortion.

[7] This discussion is taken further in M. Dahlström, 'The compleat edition'.

[8] G. Rockwell, 'What is text analysis, really?'.

provides. This is in other words a difference in perspective between *text* and *document*, manifested in the distinction between text-oriented and image-oriented editing.[9] Text-oriented editing works mostly with text transcriptions, image-oriented editing mostly with facsimile images. To much text-oriented editing of for example intentionalist descent, text is an immaterial, abstract, ideal, copy-independent phenomenon, while to much image-oriented editing in for example sociology of texts or material philology, text is rather a material, physical, concrete, copy-dependent phenomenon. Depending on which ideal you subscribe to, the editing, its tools and the resulting edition will turn out to be very different indeed.

Finally, a significant tension has emerged between displaying the uniformity or multiformity of the edited work—what Peter Robinson has referred to as the *one text* or *the many texts*.[10] The former ideal strives for choosing or constructing single copy-texts whereas the latter ideal turns the edition rather into an archive. This is a difference between on the one hand selection and discrimination and on the other hand more or less total exhaustiveness.

Will the ideals and tasks of scholarly editing change with new media?

Some of these conflicts have been around for the entire history of scholarly editing, while others have emerged during the last decades. It is sometimes claimed that these conflicts and tensions are largely a result of the Gutenberg paradigm and that new media and the web will turn things topsy-turvy and impose a new paradigm of ideals. I think this is presumptuous. In fact, digital editing does not seem to do away with this pattern of conflicts at all, but rather accentuates some of them.

For instance, the quest for totality and more or less complete exhaustiveness within scholarly editing is considerably strengthened by digital editing. The inclusive potential of digital editions and archives such as the ability to house full-text representation of many or indeed all

---

[9] The distinction between text-based and image-based editing is treated in for example the thematic issue on image-based humanities computing, *Computers and the humanities* 36, 1 (2002). See also the sharp essay on the topic by G.T. Tanselle, 'Reproductions and scholarship'.

[10] P. Robinson, 'The one text and the many texts'.

versions of the edited work and to support the modularisation of documents into movable fragments *across varying contexts*, seems to boost the idealist strand in editing. This trend is even further supported by text encoding, where form is separated from content, and where fact is quite often conceived of as separable from interpretation.

But these notions have forerunners within printed scholarly editing as well. Printed definitive editions attempted to be matter of fact, exhaustive and final. And printed parallel and synoptic editions are attempts to accommodate versionality and inclusivity within the covers of the codex book, and Kanzog's Archiv-Ausgabe of the work of Heinrich von Kleist takes this idea even further.[11] Admittedly, those forerunners are different in scale, and digital editions such as *The William Blake Archive, The Wife of Bath's Prologue*, or *The Rossetti Hypermedia Archive* can partly be seen as attempts not only to embody but to prolong these notions and make them become real.

It is worth remembering that bibliographical collections of documents derive much of their strength not only from their inclusive but also from their exclusive mechanisms.[12] Scholarly editions in fact gain much power and status by their discriminating task and in the way they define and constitute works by excluding and being precise. This task as a constitutive statement is in some contrast to the notion of total archives and do-it-yourself editions. One might however note that in a number of current digital editing projects, an 'edition' can designate a temporary, editorially argumentative selection from a more general-purpose and comprehensive storage archive. This suggests a possible distinction between edition and archive, where the former but not the latter explicitly takes a stand. At the same time there is a counter-reaction against the 'archival' trend, pleading for a return to the editorial prerogative and to primarily monoversional edition forms.

Some researchers have observed that scholars remain faithful to printed editions and that we might even be witnessing a decline in digital editing. Steding, who devotes his dissertation on scholarly editing to making a plea for the medially superior qualities of digital scholarly editions when compared to the qualities of printed editions, does not

---

[11] K. Kanzog, *Prolegomena zu einer historisch-kritischen Ausgabe der Werke Heinrich von Kleists*.

[12] The scholarly edition is here considered as a bibliographical, referential genre. It is in that sense related to such genres as the catalogue, the reference database, the library collection, the archive, or the enumerative bibliography.

really provide a good answer as to why digital editions have not already superseded or even outdated their printed counterparts.[13] Steding and others who have commented on this circumstance concentrate their analytical efforts on technical and material qualities to explain or even predict medial evolution. If we on the other hand put more emphasis on editions as biased arguments with certain meritocratic values, we can include as one of the tasks of a scholarly edition to be a report of an accomplished scholarly labour and a status-carrying constitution. In that sense, the edition shares some of the discursive and rhetorical characteristics that historiographical and genre studies have ascribed to the scholarly journal article genre: their objectivity ideal, their aim to avoid aesthetical style, their task of both constituting and defining what is (and what is not) knowledge alongside the more recognized task of presenting tools for new knowledge. To Frohmann, the flora of scholarly journal articles is primarily a reward system in the form of a gift economy. It forms an archive of accepted and constitutive state- ments rather than an arena of current and cutting-edge research. He therefore is sceptical as to whether 'scientific articles contribute infor- mation useful in the derivation of new results.'[14] He further claims scientific 'facts' are by no means absolute. They are rather statements, and a chief function of the journal article is to stabilize these state- ments by piling up armies of footnotes in their defense. There is, I think we would agree, a similar, stabilizing socio-cultural function within the scholarly edition. Not only within the community of tex- tual scholarship, but also within literary culture in general. The fact that a work has been the object of scholarly editing is a seal that it has been raised to literary nobility and invited into the inner rooms of the literary salons. Burman is on the mark when he refers to scholarly edi- tions as the 'cathedrals' of literature.[15] Frohmann dresses up a similar thought in a sophism: 'A text does not belong to the scriptures because its content is holy; rather, its content is holy because it belongs to the scriptures.'[16]

These functions of being a scientific report, an interpretative state- ment, a constitutive and canonizing tool, even a monument, provides the scholarly edition with a particular tension. On the one hand, it

---

[13] S.A. Steding, *Computer-based scholarly editions*.
[14] B. Frohmann, *Deflating information*, 46.
[15] L. Burman, 'Det enkla valet', 85.
[16] B. Frohmann, *Deflating information*, 153.

is supposed to be dynamic and as a research tool quickly reflect new findings and scholarly development. On the other hand, there are arenas where the scholarly edition is supposed to be conservative, static and confirmative. We see this two-faced character in the way scholarly editing is marked by both being prone to change, experiment, question and discussion while at the same time being highly conservative and traditional. There is a welcoming and there is a resistance. Granted, many editors have proven eager to try out new technologies and media, but equally many—perhaps more—editors have proven unwilling to experiment and change.

So in essence, we might perhaps better understand the development of, and relation between, printed and digital editions if we emphasise this multiple task of the scholarly edition as scientific report, as meritocratic constitution and as monument, and where it is plausible that the digital edition has not yet been able to accommodate the structure of meritocratic, social and symbolic values that surround the fixed and stable printed scholarly edition.

It is further possible to assume an interplay between edition types and their users according to parameters such as economical, intellectual and social class, in the sense that a typology of editions based on for example format to some extent reflects a social distribution among user groups. Genre studies indicate that already the graphical and textual pattern of a document signals genre and that different medial and material edition types might be linked to different user group adaptations. From such a perspective, it would seem probable that a division of labour would come about between printed and digital scholarly editions. The former would then be assigned the task of constitution and of a concentrated and lucid presentation of the material in a manageable format. The latter would increasingly be thought of as the archival and referential material that has made the digital end product possible or as an extension and continuation of the constantly ongoing editorial work. And this is of course to some extent what we are witnessing within many current editing projects, for example in large Scandinavian editing projects.[17] I would argue that this is because new media types do not necessarily replace older ones, but rather bring about a change of roles, a displacement of tasks and functions for the media types concerned.

---

[17] M. Dahlström and E.S. Ore, 'Electronic text editing in Scandinavia'.

Library digitisation and transmission ideals

Let us however return to the pattern of conflicts within scholarly editing depicted in Table 1 and discussed in the text that followed. I would argue that this pattern is primarily not media specific to either printed editing or digital editing. It is rather *a general trait of textual transmission* as a cultural phenomenon. We might therefore expect to detect more or less the same pattern in other cases where textual transmission has been similarly stabilized by institutionalisation. The discussion below goes more deeply into a particular case of institutionalised transmission.

One of the things new and different with digital editing is the division of labour and media that surrounds the field of editing and connects it with neighbouring activities. I am thinking in particular of the changing relation between scholarly editing and the ongoing digitisation within libraries. And interestingly enough, digitisation within libraries is developing a pattern of conflicts and tensions between ideals and transmission strategies that is in many ways similar to the pattern within scholarly editing.

Libraries and other so-called memory institutions have throughout history developed a range of methods and tools for transmitting full texts between material carriers and across media family borders. In this sense, library digitisation belongs to the same tradition as twentieth-century microfilming and the transcribing of manuscripts performed by ancient libraries and medieval copyists. The Gutenberg era marked a sharp decline in this full-text transmitting business, and libraries devoted their time to producing bibliographical meta-labels for documents rather than reproducing the full documents themselves. With digital reproduction technologies however, libraries have drawn a historical circle. They are yet again dedicating much energy and attention to the full-text transmission they largely abandoned at the dawn of the print age. In so doing, they take on a much more explicit role of producing and shaping the digitised cultural heritage in addition to their accustomed role of preserving it and making it available.

Digitisation strategies

Let us bear in mind that digitisation within libraries is much more than the mere technical capture of some content in analogue documents.

It is rather a large and quite complex chain of affairs, from for example planning, budgeting and selection, via content capture, metadata production and publishing, over to documentation, marketing and archival maintenance. The links in the chain overlap, cooperate and support one another. In principle, every link is a factor that might affect and delimit the nature and quality of the final digital resource. This includes to what extent, at what level and in what form the users are granted access to the resource. How the different links are implemented and work together is of course dependent on the overall strategy for the digitisation project.

For instance, library digitisation works with two modal strategies: *text digitisation* and *image digitisation*—similarly to scholarly editing, as we recall. In text digitisation, documents are primarily interesting as carriers of the linguistic text rather than as graphical and material artefacts. So the task is to create a machine-readable and (usually XML-compatible) encoded transcription of the text. Image digitisation on the other hand wants to capture the source documents as two-dimensional images (digital facsimiles), using scanning or digital camera. Needless to say perhaps, the two approaches are often combined.

There are other distinctions of digitisation approaches around as well, such as that between proactive (i.e. just-in-case) and reactive (i.e. just-in-time) digitisation, or between conservational (non-tampering) and restaurational (tampering) digitisation. One will quite quickly see similar distinctions within scholarly editing as well.

Mass digitisation

But the currently most spoken-of strategy is, no doubt, mass digitisation. It aims to digitise massive amounts of documents (thus an all-inclusive strategy) using automated means, in a relatively short period of time,[18] such as Google Book Search, Europeana, Open Content Alliance, the Norwegian DigitALT or the late Microsoft Book Search. It operates on an industrial scale and with as many chain links as possible fully automated. Mass digitisation systematically digitises whole large collections, document by document, with no particular means of discrimination. The projects might assume more or less ambitious

---

[18] K. Coyle, 'Mass digitization of books'.

totality claims: from projects limited—by copyright, politics or admin-istration—to a particular subset of a collection to the grand supercol-lection schemas mentioned above. The idea is to digitise 'everything' within the collection or sets of collections.

For practical reasons, mass digitisation has to minimize manual and labour-intensive work and cannot include intellectual aspects such as textual ambiguity, interpretation, descriptive text encoding and man-ual proof reading. Neither can it afford to have too much metadata and information about the source document accompanying the digital representation. In mass digitisation, transmission has been flattened out into a linear streamlined affair. Mass digitisation has its ambitions and its value in *scale*, not in depth. Vast numbers of books are made available and their texts searchable. Projects such as these have been met with no shortage of critical remarks. From a bibliographical and archival standpoint, Google Book Search has for example been criti-cised by Duguid on the grounds that:[19]

- It produces representations with poor textual and graphical quality;
- It supplies scarce and occasionally erroneous bibliographical infor-mation about the source documents as well as the digital files;
- It does not seem to pay particular attention as to which edition or version to use as source, and,
- It is not particularly transparent what will happen with the digital material in regard to, for example,
  ○ Intellectual property rights (such as who will in the future be con-sidered as holder of copyright and the right to edit the digital files?);
  ○ Economy (for which services will the future end-user likely have to pay a fee?);
  ○ Preservation (who will assume administrative and technical responsibility for the long-term curation of the digital material?).

One might also question whether the libraries involved in Google Book Search will be able to turn to other digitising agents with the same source documents in the future, when Google's project has come to an end or if Google should cease altogether as an enterprise.[20]

---

[19] P. Duguid, 'Inheritance and loss?'.
[20] Let us bear in mind that Google Book Search is a project limited in time, and not an open-ended activity.

In spite of these critical remarks one can definitely see strengths and positive effects of mass digitisation projects. They combine on the one hand commercial agents who are strong in financial resources but in need of contents, with on the other hand public libraries who are reversely strong in contents but in need of financial resources. A *marriage made in heaven*, it would seem. The result: a gigantic, growing bank of digital texts that can be searched free-text, used as localising tool, and—perhaps more importantly—form the technical base for many kinds of future software development and implementation.

But mass digitisation is not the only strategy around. Only a limited number of large libraries have the interest, competence and resources to implement it. It further suits some objects and collections and not others, such as fragile books, manuscripts and other unique objects, and documents whose texts are difficult to read or interpret. Cases such as those require considerable resource-consuming and manual labour, and can not reasonably be referred to as mass digitisation. We find in fact a number of digitisation projects in libraries world wide that are anything but 'mass'.[21] What then do we call them? I suggest we refer to them as *critical* digitisation.

Critical digitisation

Critical digitisation implements several of the links in the long digitisation chain in a manual, intellectual, critical way. At every step one can make choices, deselect and interpret. Mass digitisation turns a blind eye to most of these choices, whereas critical digitisation acknowledges and makes active use of them. The project might for example focus on a single document or a limited set of documents. It may need to perform a deliberate and strategic selection from a number of possible and more or less complete source documents. Perhaps the source document has text that is difficult to decipher and decode. The text or image may need to be edited and manipulated to make sense or context. Perhaps it is vital not to destroy the source document during the digitisation process (as mass digitisation does) but rather subject it to careful preservational or conservational measures. The project may

---

[21] Two Scandinavian examples are the Codex gigas (http://www.kb.se/codex-gigas/eng/) at the National Library in Sweden and the Dirik family scrapbook at the National Library in Norway (http://www.nb.no/baser/diriks/virtuellealbum/index.html).

wish to manually and critically produce a representation that is as faithful and exhaustive *as possible* in relation to the source document and its text, its graphics and perhaps artefactual materiality. The digital object may need to be provided with large amounts of metadata, indexing, descriptive encoding, paratexts and bibliographical information, i.e. bibliographical and other scholarly research may need to be *embedded* in the objects themselves. Critical digitisation is qualitative (or idiographic) in the sense that it concentrates on what is unique and contingent in the documents, whereas mass digitisation is quantitative in its design to capture what are common, regular, foreseeable traits in large amounts of documents (i.e. nomothetic). In consequence then, critical digitisation normally has to develop project-contingent procedures and tools and tailor them to the nature of the documents in the particular collection. In mass digitisation, the single documents in the digitised collection are on the contrary subordinated (or 'tailored') to more general, perhaps even universal procedures and tools.

Critical digitisation is in other words a more exclusive strategy—in more senses than one. Let us briefly sum up the differences between the two approaches using Table 2 below (and bearing in mind also Table 1).

Table 2: Tensions in library digitisation: critical versus mass digitisation

Critical digitisation	Mass digitisation
is primarily manual	is primarily automated
critically recognizes the distortion digitisation brings about	in effect treats digitisation as a cloning process
undertakes a well-informed selective analysis of source copies	normally picks whatever source copy praxis or chance happens to present
maximises interpretation and metadata	minimises interpretation and metadata
is idiographic	is nomothetic
uses contingent tools	uses universal tools
treats graphical and material document as artefact	treats linguistic text as fact
discriminates	is exhaustive
works in depth	works on scale

I am aware that Table 2 might be interpreted as painting a flattering picture of critical digitisation, while mass digitisation is attributed a negative role. That is however not the aim here. The advantages as well as the disadvantages of mass digitisation can be reversed in the case of critical digitisation. Critical digitisation is slow and very costly in relation to the number of produced objects. It addresses a small audience. It may require rare skills in for example textual and bibliographical scholarship. It often has an image-oriented ideal where the facsimiles are left without accompanying machine-readable transcriptions. The result risks quickly being more or less forgotten after the project is completed, and many digitisers neglect to inform about and market the project in proportion to the labour invested. Manual labour such as interpretations and tailored technical solutions are seldom properly documented (if at all), but run the risk of becoming silent knowledge locked in the minds of the digitising persons, and therefore available to the institution only as long as the persons remain employed by it. Mass digitisation on the other hand requires such an industrial scale that its strategies, practices and technologies need to be documented in order to be properly implemented by many different people and machines over long periods of time.

There is also another consequence of the way critical digitisation intervenes in the documents. Someone visiting a critically digitised collection faces a material that is in a sense already encoded, manipulated, labelled and explicitly interpreted. The more this has been done explicitly, the more it turns into a sort of *comment* on the source document which might work counter to how flexibly the material can be reused in new contexts. Mass digitisation on the other hand conveys an aura of objectivity around the objects and a lack of manipulation— but that is of course due to the cloning ideal of mass digitisation (an objectivity that is in effect nothing more than a chimera).

So opinions as to how usable and reusable the products of different digitisation projects really are to scholars and scientists can radically differ. On the one hand, critical digitisation enriches its objects with an intellectual added value and applies some kind of quality seal as regards selection, textual quality, resolution, proof-reading, comments and bibliographical information. On the other hand, not all scholars may be interested in that particular metainformation and added value, but are in need of quite different aspects than those that happened to catch the interest of the digitising institution.

One can even turn it all around and claim that critical digitisation risks falling for another kind of cloning ideal than the one previously identified as typical for mass digitisation. That is to say, this other

cloning ideal might work on the assumption that as long as the digitisation process inscribes in the digital representations large and deep
enough metainformation, we will obviate any future need of new
digitisation efforts, since all material and all possible aspects already
exist in the digital archive we have created. We would in other words
be facing the 'definitive' digital representation, *once and for all.* Mass
digitisation on the contrary might be thought of as more advantageous
precisely because it does *not* select, provide metadata and explicit text
encoding and interpretation, but rather constructs reservoirs of source
documents that scholars ideally can use, reuse and enrich the way it
suits them best. We obviously recognize this pattern from scholarly
editing and its tension between Alexandrian and Pergamanian ideals. Again, however, we need to remember that the products of mass
digitisation can as well be thought of as dependent on interpretations and selections—but that these are in effect ignored and silenced,
which leaves the user helplessly dependent on the unknown choices
that praxis forced upon the mass digitising institution. We should also
readily admit that the lack of textual quality and proof-reading in for
example Google Book Search hardly makes any scholar particularly
happy, regardless of his or her disciplinary affiliation.

It would seem clear that both strategies have their advantages and
disadvantages to different audiences. Whether a digitising library
adheres to an ideal that is closer to critical digitisation or to mass
digitisation, however, should take into account which kind of digital
material is being produced, what meta information and added values
should be attached to the material, to what extent it should be able to
be used and reused, and to which user group it should prove to be of
interest. In that way the chosen digitisation strategy legitimises certain
kinds and levels of material at the expense of others, and favours certain user groups over others—all a question of symbolic power. There
are of course mechanisms in digitisation strategies that might be des-
ignated as constitutive and perhaps canonising. Again, we recognise
this tension from our previous discussion about the socio-cultural
functions of scholarly editing. All in all, however, the library community in general is increasingly favouring the ideals of mass digitisation
and its pragmatic notion of transmission as a relatively simple, linear, content-capturing affair. The critical digitisation activity is much
smaller and is in many cases currently threatened to become extinct
as an ineffective, over-costly luxury undertaking.

And the library community probably *is* more suitable to be engaged in mass digitisation. Kjellman observes that whereas museums discriminate, select and deselect as part of their joint collecting task, national libraries in particular display an ambition of comprehensiveness in their collecting activity.[22] Historically this has created an ideal of objectivity within the national library communities that tends to hide the discriminating mechanisms of the institution. Such an ideal is certainly expressed and fuelled by mass digitisation. The library collections, furthermore, are largely made up of printed documents that are mass produced to begin with. Given that the many copies of a published book are normally thought of as identical, it is consequently thought of as more or less indifferent whether the one copy or the other is picked (rather than selected) as source document, i.e. as 'ideal copy'. Archives and museums on the contrary manage unique objects to a much higher degree, and their digitisation in consequence regularly concerns single document artefacts rather than the text as a presumed commonality in for example a book edition.

Scholarly editing and library digitisation

So to some extent, the area of library digitisation seems to develop into a map of tensions and conflicting ideals (Table 2) that bear considerable intellectual similarities to that of scholarly editing (Table 1). As scholarly editing does, library digitisation can make deliberate selections and discriminations, interpret, analytically compare source document candidates and seek to establish something common, sometimes even ideal, in such a collection of candidates. They both edit, optimise, document, comment and produce metatexts. They are both examples of transmission practices that are stabilised and legitimised by socio-cultural prestige institutions and that are based on agreed upon, publicly declared and fairly documented principles. In their exclusive character, they both constitute, consecrate and perhaps even canonise works of culture. If scholarly editions are cathedrals of literature, one might certainly argue that ambitious critical digitisation projects within libraries turn the selected documents into national monuments—or even testaments.

---

[22] U. Kjellman, *Från kungaporträtt till läsketikett*, 239 f.

I would contend that in the case of critical digitisation, libraries are in effect engaged in what comes close to textual criticism. Librarians and other employees involved in current digitisation projects might feel awkward with such a label of textual criticism or even bibliography put on their work, but the historical connection cannot be denied.

The major *distinction* one might arguably define between scholarly editing and library digitisation concerns the establishing of the text: library digitisation does not seek, as scholarly editing does, to establish a text (a copy-text as it were) that perhaps never existed previously and which cannot be literally transmitted from a single source document. But this argument can be countered—in two ways.

Firstly, more and more scholarly editions are constructing documentary archives by minimising editorial interventions and instead providing fulltext and 'raw' versions, at times even refraining from highlighting or constructing a single uniform established base text. Secondly, critical digitisation increasingly approaches scholarly editing, for example in image-oriented digitisation of documents where we have more than one copy to choose from. There, the critical comparison of several source candidates based on their condition, completeness etc. results either in the deliberate selection of one candidate or in an eclectic amalgam of fragments from several candidates (such as eclectic facsimiles). In the latter case—at the very least—we are facing, if not the ambition of textual criticism to establish an ideal text, then at least a kind of document criticism that seeks to establish an ideal document.

Besides, there is a quite tangible cooperation between the two fields. Libraries normally house the source documents of interest to scholarly editors to begin with, and often perform the technical digitisation of them to serve large editing projects. And libraries are arguably best suited to be responsible for the long-term technical and bibliographical maintenance and preservation of the digital files. They are also in a good position to coordinate and manage the intricate web of IPR interests within large editing projects in a way other agents, including scholarly editors, simply cannot do. This is particularly the case with image oriented projects, where libraries and archives often *are* the very IPR holders themselves.

Increasingly, furthermore, scholarly editors are implementing practices, systems and tools that were developed by the library community for managing, relating and describing large collections of documents, such as the Metadata Encoding and Transmission Standard (METS) for metadata and the Functional Requirements for Bibliographic Records (FRBR) for descriptive cataloguing. Library digitisation projects are also valuable to textual scholars and scholarly editors in the way they make large amounts of material available, if nothing else as facsimiles that can to an increasing degree be subjected to OCR and whose text therefore can be turned machine-readable and thus reusable in scholarly editing projects. Not to mention the many digitised manuscripts that have previously been unpublished and not subjected to research but that are now becoming identified, catalogued and made available. Again, the library's policy for copyright, accessibility and technical quality becomes an absolutely crucial factor in determining the usability and reusability of the material for scholarly research.

So scholarly editing and library digitisation are perhaps approaching a point where the two not only meet but even merge, at least on the project level. What happens in effect when scholarly editions and archives more and more turn into digital libraries, and what happens when critically digitised collections within libraries become increasingly granulated, technically and architecturally sophisticated, thus increasingly take on the form of 'editions'?

Scholarly editors will likely need to have access to documents that are information dense, carefully prepared and proof-read, enriched with bibliographical information, and that are accessible and editable in high resolution or deep encoding. Such needs are obviously better met by critical digitisation than by mass digitisation. Libraries on the other hand might be expected to either only produce information-poor files (in the mass digitisation mode) or else (in the critical digitisation mode) restrict access to or usage of the information-rich files—which is in fact increasingly the case (due to legal, economical, or administrative reasons) and which poses a major problem for future research and reusability.

At the same time, if the current trend within libraries towards mass digitisation turns paradigmatic, and if scholarly editing increasingly will cooperate with those libraries, we will likely see a boosting of the idealistic, positivistic strand, where source documents and their texts are treated as facts rather than as expressions of subjective statements. Whether this is a good or a bad thing remains to be seen.

References

Bazerman, C., *Shaping written knowledge. The genre and activity of the experimental article in science* (Madison, WI, University of Wisconsin Press, 1988).

Bodard, G. and J. Garcés, 'Open source critical editions. A rationale', in *Text editing, print, and the digital world*, eds M. Deegan and K. Sutherland (Aldershot: Ashgate, 2009), 83–98.

Burman, L., 'Det enkla valet. Konsekvenserna av en oproblematisk textsituation'. In *Vid texternas vägskäl: Textkritiska uppsatser*, eds L. Burman and B. Ståhle Sjönell (Stockholm: Svenska vitterhetssamfundet, 1999), 83–99.

Coyle, K., 'Mass digitization of books', *Journal of academic librarianship* 32, 6 (2006), 641–645.

Dahlström, M., 'The compleat edition', in *Text editing, print, and the digital world*, eds M. Deegan and K. Sutherland (Aldershot: Ashgate, 2009), 27–44.

Dahlström, M. and E.S. Ore, 'Electronic text editing in Scandinavia', in *Bausteine zur Geschichte der Edition: Skandinavische Editionen*, eds P. Henrikson and C. Janss (Tübingen: Niemeyer, forthcoming).

Duguid, P., 'Inheritance and loss? A brief survey of Google Books', *First Monday* 12, 8 (2007), viewed on 14 January 2009, http://firstmonday.org/htbin/cgiwrap/bin/ojs/index.php/fm/article/view/1972/1847.

Frohmann, B. *Deflating information. From science studies to documentation* (Toronto: University of Toronto Press, 2004).

Greetham, D., 'Textual scholarship', in *Introduction to scholarship in modern languages and literatures*, ed. J. Gibaldi (New York: MLA, 1992), 103–137.

——, *Theories of the text* (Oxford: Oxford University Press, 1999).

Kanzog, K., *Prolegomena zu einer historisch-kritischen Ausgabe der Werke Heinrich von Kleists. Theorie und Praxis einer modernen Klassiker-Edition* (München: Hanser, 1970).

Kjellman, U., *Från kungaporträtt till läsketikett. En domänanalytisk studie över Kungl. bibliotekets bildsamling med särskild inriktning mot katalogiserings- och indexeringsfrågor* (Uppsala: Uppsala university, 2006).

Renear, A., 'Literal transcription. Can the text ontologist help?'. In *New media and the humanities. Research and applications*, eds D. Fiormonte and J. Usher (Oxford: Oxford university, 2001), 23–30.

Robinson, P., 'The one text and the many texts', *Literary and linguistic computing* 15, 1 (2000), 5–14.

Rockwell, G., 'What is text analysis, really?', *Literary and linguistic computing* 18, 2 (2003), 209–219.

Steding, S.A., *Computer-based scholarly editions. Context, concept, creation, clientele* (Berlin: Logos, 2002).

Tanselle, G.T., 'Reproductions and scholarship', *Studies in bibliography* 42 (1989), 25–34.

THE POSSIBILITY OF SYSTEMATIC EMENDATION

John Lavagnino

This essay reconsiders the practice of textual emendation in the light of corpus methods. Because traditional methods are based in part on linguistic knowledge and comparison with other texts, there is a clear role for new methods that can automate the collection and evaluation of large amounts of relevant data. But there is also a dilemma, because emendations are of two types: sometimes we change the text to a more likely reading, sometimes we change it to an unusual, perhaps unprecedented reading. Automating the decision about which direction is necessary at a particular point in a text remains a problem.

Introduction

My aim is to explain the traditional practice of emendation by scholarly editors—what it is and why we do it, and some positions for and against it—and then to reconsider the practice in the light of corpus methods; by which I mean approaches, like those in other contributions to this book, that ultimately depend on having very large amounts of text to draw on in a mechanized way. I know this definition is not perfect for a lot of what people are actually doing, and even then there is good work that is not heading in this direction at all—but it does at least distinguish this kind of approach from more traditional ways. My aim is to talk in general terms about the possibilities for digital emendation, rather than to discuss specific techniques; what this will establish is a modestly revised idea of what emendation is, rather than a dramatically new method. I do not have any new emendations of my own to reveal, but I do hope to show that even in this area, traditionally considered to require the greatest personal knowledge of the language and literature in question, there is a significant role for corpus methods.[1]

---

[1] Elsewhere in this collection Ben Salemans discusses emendation in the context of stemma-building, since one reason to build a stemma is to support work on emendation. But of course there are reasons to try to build a stemma even if you do not plan to try to make emendations: it tells you something about the history of the text; and

Text emendation

What is the justification behind emendation—behind saying that we believe our text is wrong and that we know better, and accordingly will alter the text we offer to readers?

One thing we know about textual transmission is that a copy made by a human being—writing, or typing, or setting type—will frequently introduce changes. No transcription of a text of any length that has not been checked is likely to be without error. Matthew Spencer and his colleagues asked postgraduates to copy texts for them, in order to obtain a complete network of copied manuscripts to study. They of course found numerous divergences from copy to copy, at a rate close to that found in some medieval manuscript traditions.[2] Roberto Busa's digital text of the works of St Thomas Aquinas, one of the first large digital texts, went through three stages of checking, that still left it with 1600 errors, among them the omission of one entire line.[3] One new argument in the edition of Thomas Middleton's works that I helped edit was Gary Taylor's demonstration that Middleton was an inaccurate copyist of his own writing: some manuscripts of *A Game at Chess* were written by him, but they are among the worst witnesses to the correct text of the play.[4] Good copying is attainable: it is possible to have a very low error rate even in verbal recollection, but only if you learn to work very precisely indeed.[5]

Many such errors are not difficult to remedy. Any attentive reader of texts will routinely observe misspellings, faults in punctuation and other errors and will make them right with little thought. In fact, it is precisely our facility to know what was meant rather than what has been written that contributes to the persistence of obvious error in texts. In one of the explanatory notes of our edition of Middleton's works, 'disease' is misspelled as 'desease': just as with Busa's texts, repeated scrutiny by several people still failed to catch this.[6] Common forms of this problem are simple enough that they are routinely corrected nowadays by mechanical means. The spelling checkers in word

---

conversely you may want to emend a text that only exists in one witness, so that no evidence at all from comparison with other witnesses is available.

[2] M. Spencer et al., 'Phylogenetics of artificial manuscripts', 505.
[3] R. Busa, 'The annals of humanities computing', 87.
[4] G. Taylor, 'Praestat difficilior lectio'.
[5] M. Calinescu, 'Orality in literacy', 176–177.
[6] G. Taylor, 'Praestat difficilior lectio', 114.

processors often fall down on proper names, but better systems do not: Google is able to correct many misspellings of my last name, for example, even though it is obviously not all that common even in its correct form. The correct form is still much more common online than the incorrect form and so provides a basis for the Google engine's guess.

The emendation debate

But emendation is a laborious and inevitably speculative process, and it has numerous critics. For many texts of the last two or three centuries emendation does not seem that important an activity. The prestige of good emendation is associated with classical scholarship, and so with a body of work that has come down to us through a number of copyings and recopyings and is often in evident need of repair. The case is very often otherwise with more recent works. A modern scholarly edition by Fredson Bowers of *The Scarlet Letter*, a novel that is over 80,000 words in length, introduced eleven changes in the spelling of words (to bring them into line with Hawthorne's normal practice), nine in punctuation, five in the category of 'obvious typos'—changes that in fact had been corrected by the original publishers when they reprinted the book—and two actual new corrections, one a correction so slight that few would notice, leaving a single correction that involves supplying a missing word.[7] Nineteenth-century editions of the novel say:

> It is essential, in order to a complete estimate of the advantages of official life, to view the incumbent at the in-coming of a hostile administration.[8]

A verb seems to be needed after 'in order to'; Bowers supplied 'form'. Other verbs would also work, and Bowers did not discuss his choice—which however echoes the language of the Preamble to the Constitution of the United States of America.

The change seems necessary, but it also seems trivial. The contribution of Bowers's edition is much more in tracing the history of the text and establishing that it is indeed reliable, than in introducing emendations. The textual situation of many recent works is more complicated, but there are also many that are comparable to *The Scarlet Letter*. Donald Pizer

---

[7] N. Hawthorne, *The scarlet letter*, 1962.
[8] N. Hawthorne, *The scarlet letter*, 1850, 48.

and Jerome McGann have argued the point in more detail: that perhaps emendation is just not that important a function of the editor
when dealing with texts of the modern copyright era, when author and
publisher have a collaborative relationship.[9]

Emendation may not be especially necessary; it may also work
against what scholars using an edition actually want. The linguist
Roger Lass takes that position:

> One obvious motivation for not emending is that emendation is not in
> fact creative, but destructive; it obliterates part of the record, and substi-
> tutes for it an invention of another time, place and culture. Rather than
> filling an epistemic gap, as reconstruction or interpretation of docu-
> ments do, it falsifies the record, and produces second-order witnesses
> not in essence principle-driven, but dependent only on argument (often
> flabby and aesthetic rather than linguistically controlled).[10]

In textual-editing circles, the word is 'contamination'. Complaints
are common about scribes who changed things they were copying—
but emending editors are doing that too. As G. Thomas Tanselle put
it, there is a difference 'between transcription (where the aim is to
make no alterations) and critical editing (where the aim is to introduce alterations)'.[11] But historical linguists have no trouble producing
examples of ways in which, emendation apart, even modernization of
spelling and punctuation can distort meaning (see Norman Blake);[12]
Tanselle himself is opposed to modernization, for that reason among
others, but it is still a common practice among editors of English-
language texts before 1800.

Emendation may appear either unnecessary or anti-historical; it can
also appear inconsequential. Is *The Scarlet Letter* all that different even
if we read a reprint with an obvious error caused by bad OCR every
five pages? In the era of large corpora, one inclination is to stop worrying about such details, because it is the large bulk of text that really
matters. But even from the point of view of the traditional reader we
could question whether every word needs to be right. As Jorge Luis
Borges commented:

---

[9] D. Pizer, 'On the editing of modern American texts'; J. McGann, *A critique of
modern textual criticsm*, 75.

[10] R. Lass, *Historical linguistics and language change*, 100.

[11] G.T. Tanselle, Untitled review, 488.

[12] N. Blake, *A grammar of Shakespeare's language*.

The perfect page, the page in which no word can be altered without harm, is the most precarious of all. Changes in language erase shades of meaning, and the 'perfect' page is precisely the one that consists of those delicate fringes that are so easily worn away. On the contrary, the page that becomes immortal can traverse the fire of typographical errors, approximate translations, and inattentive or erroneous readings without losing its soul in the process.[13]

This is an argument for the primacy of a copious author like Cervantes, whose genius is more in evidence in bulk rather than in the single page; and generally for the primacy of large forms rather than lyrics. It has often been observed that national differences in editorial approaches stem partly from the differences in the textual situations of classic authors: Homer, Cervantes, Shakespeare, and Goethe variously lead you to develop quite different models of what texts are normally going to be like, in terms of the reliability of the sources and the amount of emendation that will be called for. But even then, of course, readers and scholars constantly find matters to interest them precisely in single pages; and even editions of prose works do want to eliminate mistakes that might cause readers to stumble and get held up wondering.

Against these arguments that emendation is unnecessary, dubious and inconsequential there is first of all an anthropological argument. People notice mistakes in texts, want them fixed, are puzzled if they are not. If we are like scribes in that we want to fix things that seem wrong, perhaps we need to accept the impulse and try to regulate it, rather than imagining we could halt it entirely. In other words, it seems *strange* to leave errors in place. And as a practical matter, it is very difficult to get a text reprinted exactly, errors and all, because of that same impulse to correct: but once you resolve to fix the obvious errors you then need to accept responsibility for listing and defending them. Publications that claim to be exact reprints that nevertheless introduce changes are common but unscholarly. The one route that is practically possible in this direction is the photographic reprint.

A response to Lass and other linguists would be to make better-informed emendations: these complaints are in part about emendation done without a good knowledge of the history of the language behind them and often they seem to relate to rather old editions, or the worst

---

[13] J.L. Borges, 'The superstitious ethics of the reader', 54.

of what editions have to show. George Kane has also pointed out that there is real force, precisely from the historical point of view, to an argument favoring emendation:[14] things which are wrong in original documents, and that would have been regarded as obviously wrong by readers in the period, are *also* not to be regarded as good historical evidence, any more than badly emended texts would be. My 'desease' example should not be interpreted as evidence that such a spelling was widely accepted in our age or by our editorial team. Once we have observed that simple error does occur when texts are copied, we have to conclude that even a perfect copy of an erroneous text will reproduce some errors.

Emendation and corpus methods

Corpus methods are not distant from the text and reliant only on the general drift, on what Borges' comment sees as the real soul of the text: one source of their great power is the way they work with the words that are there rather than anything else. The consequence is that we do want to be working with the right words. And certainly readers have felt that obvious errors are consequential: the minimum level of error in a reading text that is necessary to make readers doubt the text is quite low, much lower than current error rates for uncorrected OCR on most material. So in this alternative vision emendation is not unnecessary but unavoidable, not anti-historical but essential for correct historical data, not inconsequential but a vital protector of the apparent integrity of the text.[15]

Those reasons also suggest that we should think about how emendation might benefit from corpus methods. But there is a more general argument in favour of corpus-driven emendation: if you have a new and powerful method in the humanities you need to either show how it changes established and valued practices, or why it is still powerful even though it has no bearing on them. In the sciences it makes sense to claim that something new and important might only affect a niche; indeed much of the time nowadays innovation comes in highly circumscribed areas. And in the sciences one also frequently depends

---

[14] G. Kane, 'Conjectural emendation'.
[15] See F. Kermode, 'Divination', and G. Taylor, 'Inventing Shakespeare', for some further reflections on the importance of emendation.

on using *results* delivered by another subfield; the chemist can use bits of physics that come up without worrying about their grounding. This is not the case in the humanities, where we feel that things are much more intertwined and that as a literary scholar I am engaging in risky behavior by just *using* historical or linguistic results; I need to know how these results were established. So if corpus methods are taken to really matter, they alter the very nature of text emendation; but if corpora leave a lot of ground unaffected, there is some reason to rethink the case for these methods.

My Google example shows that there is of course a very clear contribution to emendation from corpus methods, that is, the familiar contribution of spell-checking. In the Google case the possibility of emendation rests on a vast, untagged corpus rather than on the combination of a dictionary and an analysis of common spelling errors, as has been the usual approach in word processors for some decades now. But I hope to show that corpus methods motivate us to rethink emendation, to observe that the work of emendation really points in two directions, at different times and with different texts: towards looking for *normal* or *conventional* language on the one hand, and towards looking for *unique* or *distinctive* language on the other. One could call these two cases the emendation of prose and of poetry respectively, but that cuts too many corners.

We have seen a case in relation to Hawthorne's prose (*The scarlet letter*) of an obvious error that is also readily fixed, though perhaps not in exactly the way the author would have had in mind. Here are two further relevant examples of emendation. The first is from a famous passage in Shakespeare's *Life of Henry the Fifth* describing the death of Falstaff:

> for after I saw him fumble with the Sheets, and play with Flowers, and smile upon his fingers end, I knew there was but one way: for his Nose was as sharpe as a Pen, and a Table of greene fields.[16]

It has long been felt that there is an obvious error here, but one without an obvious emendation. Many editions since the eighteenth century have changed 'a Table of greene fields' to 'a [= he] babbled of green fields'; not all editions even today make this choice, and there is no certainty that it is correct, but it is representative of one kind of

---

[16] W. Shakespeare, *Mr William Shakespeares comedies*, 1623, h4r.

emendation that is valued particularly highly, in which puzzling words
are turned into something that is effective and even seems typical of
the author's style.

My second example is of an emendation that creates something
typical of the author's style, but starting from words that are not puzzling at all. A.E. Housman read some lines by Walter de la Mare in a
newspaper and knew something was wrong:

> Oh, when this my dust surrenders
> Hand, foot, lip, to dust again,
> May these loved and loving faces
>   Please other men!
> May the rustling harvest hedgerow
> Still the Traveller's Joy entwine,
> And as happy children gather
>   Posies once mine.[17]

There is no flaw in grammar or sense in this stanza, but Housman
claims that he knew at once that 'rustling' was wrong, and that he
found the right word in short order.[18] In his published account Housman does not state the right word, but de la Mare's books have 'rusting'. Yet if de la Mare's poem had never survived in another text, we
would be unlikely to regard Housman's emendation as certain: even
if 'rusting' is the better word, de la Mare might just have written the
worse one at this point.

What can corpora do for us in these three cases? In the Hawthorne
case, we might look to see what phrases beginning 'in order to...' are
common in Hawthorne's writing and in the writing of his era, to consider whether some other emendation might be better than 'in order to
form'. But the result is that we find this is almost certainly a miscorrection.
Simple searches on appropriate corpora, or even just on Google Books,
produce many instances of the verbless expression; it is simply an idiom
that is no longer part of the language, one that was more common at
the time in legal and religious texts, but that also appears in Herman
Melville's *Mardi*. The case is one that supports Roger Lass's position:
the scholarly edition actually distorts the real historical form. All of

---

[17] A.E. Housman, *Selected prose*, 74.
[18] *Ibid.*, 52.

this was actually pointed out long ago, by Hershel Parker and Bruce Bebb: in the case of intensely studied writers corpus searches have often revealed more evidence for proposals made long ago, rather than producing entirely new readings.[19] That is the case with several recent articles on difficult passages in Shakespeare, which do not refer to the use of corpora but show some signs of drawing on them all the same.[20] In the 'Table of greene fields' case, nobody has turned up anything new using corpora to support the original reading. We may find that the original phrase is comparatively rare. Except that in any general collection like Google Books it appears rather frequently now because, just as I am doing here, scholars have been discussing it. Similarly, the data for 'a babbled of green fields' is distorted by prior attention.

The de la Mare example can be pursued further: Google Books offers large amounts of relevant text, though here again discussion of this very case accounts for many instances of 'rustling harvest' and 'rusting harvest'. But even if you eliminate all the direct references, the finding is that the corpus still disagrees with Housman: other poets were quite happy to say 'rustling harvest', and only a few say 'rusting harvest'. They may of course be referring to de la Mare: one clear finding from the exercise is that this poem by de la Mare, *Fare Well*, was once much more widely read than you would expect on the basis of his twenty-first-century reputation. A corpus emptied of influence by this intervening factor could support either reading, but a majority vote of collocations would still favor the wrong reading, not the right one. An algorithm that looked at more of the context ought to work better: entwining may be more associated with rusting than with rustling, for example. Yet the corpus-based analysis remains worrisome, since the plain frequency evidence suggests that de la Mare was intentionally doing something out of the ordinary: that his explicit choice was for the *unusual* word (rusting), and the misprint (rustling) was the ordinary choice, ordinary enough that Housman did not believe de la Mare had written it.

Housman's account illustrates two steps of emendation: detection and correction. But detection seems to involve two different operations. In *Henry V*, an unusual and puzzling phrase suggests error:

---

[19] H. Parker and B. Bebb, 'The CEAA', 136.
[20] T. Billings, 'Caterwauling Cataians'; T. Billings, 'Squashing the "shard-borne beetle" crux'; G.B. Evans, 'The shard-borne [-born] beetle'.

emendation is called for because the text is too strange. In *Fare Well*, emendation is called for because the text is too ordinary. We can readily imagine corpus methods to locate problems of the first kind: these are an extension of the traditional approach, of reading and looking for possible problems. And as in the Hawthorne case, corpus methods may recognize the correctness of something we had wrongly judged to be strange. We can also use corpus methods to find ordinary expressions. But doing both seems to involve us in doing two contradictory things.

And similarly, correction faces in two directions. In the 'desease' case discussed earlier, an unusual or unheard-of form, one that appears in a corpus at very low frequency, should be replaced by a very common form. In the 'rustling' case, what may be a common form, one that from the corpus point of view is not unusual, should be replaced by one that may be less common, and certainly is in this case. The traditional world of emendation shows some recognition that the practice is multiple in this way. An edition will list corrections of obvious errors but will not typically state an argument for them; yet there may equally be substantial discussion of other cases. There is a general recognition that in some situations the appropriate but more unusual reading may be preferable.[21] Choosing one approach or the other as the case seems to demand is easier for a person than for a mechanical process, precisely because in either case a motivation is still needed in support of the emendation process. So we seem to be left in a situation where the traditional approach to emending texts is supported by new data but is not easy to automate further. We can detect some locations of potential error and we can discover some possible corrections; but we will get less support when the text is being more original, is specifically doing things that we might value particularly highly.

New directions in systematic emendation

What we can do well already is the spelling-correction class of problems, which is not to suggest that this is an unimportant class as it clearly accounts for many instances. This approach has been extended to matters of fact: Daniel J. Cohen and Roy Rosenzweig developed a

---

[21] As articulated in G. Taylor's 'Praestat difficilior lectio'.

program, H-Bot, that will answer questions like 'When was Thomas Middleton born?', solely by searching for web pages that state this fact and collating the answers.[22] This particular question is often answered incorrectly in twentieth-century reference books. The nineteenth-century edition of the *Dictionary of National Biography* published the guess that was then current, namely 1570; and although Mark Eccles published his discovery that Middleton was baptized in 1580 in 1931,[23] many people clearly continued to rely on the old *DNB* date, right up to the publication of the thoroughly revised version in the twenty-first century. But H-Bot answers the question correctly.

The program is not trying to do anything especially clever: instead, it is extracting an answer from a body of data very like that created by a group of manuscript copyists. Independent people are creating web pages and copying dates from other sources; mostly they get it right. When new sources provide a new date, that trickles into the tradition and starts to spread, and no doubt some copyists notice that several dates circulate and try to find out the right one: as Cohen and Rosenzweig observe, it is because of this effect that H-Bot is surprisingly good at tracking new scholarly findings. The H-Bot experiment suggests that there is a good chance of using corpus techniques on another sort of emendation: in cases where there is choice from among possibilities already present in texts in the tradition. 'Rustling harvest' seems to have some tendency to occur online as an erroneous version of de la Mare's own text—through OCR error, mistaken transcription, or erroneous emendation?—but there are also plenty of copies that preserve his correct reading. Even a simple majority vote would produce that correct reading from the corpus. In this case, where we are comparing similar texts within the corpus, rather than trying to detect error and possibly generate a new reading not itself present in the corpus, we are much more in control of the information we need to do the job.

In summary, then: it is the discovery of one type of error (that gives us a plausible but wrong reading) and of one type of emendation (that gives us an unusual but right reading) that corpus methods are least promising for. The strikingly new emendation has long been seen as resulting only from exceptional intuition and knowledge, and as such it is not so surprising that emendation might be difficult ground for corpus methods.

---

[22] D.J. Cohen and R. Rosenzweig, 'Web of lies?'.
[23] M. Eccles, 'Middleton's birth and education'.

But important parts of the emendation terrain are amenable to corpus methods, and we can hope for further advances.

References

Billings, Timothy, 'Caterwauling Cataians. The genealogy of a gloss', *Shakespeare quarterly* 54,1 (Spring 2003), 1–28.
——, 'Squashing the "shard-borne Beetle" crux. A hard case with a few pat readings', *Shakespeare quarterly* 56, 4 (Winter 2005), 434–447.
Blake, Norman, *A grammar of Shakespeare's language* (Basingstoke: Palgrave, 2002).
Borges, Jorge Luis, 'The superstitious ethics of the reader', translated by Suzanne Jill Levine, in *The total library. Non-fiction 1922–1986*, ed. Eliot Weinberger (London: Allen Lane The Penguin Press, 2000), 52–55.
Busa, Roberto, 'The annals of humanities computing. The *Index Thomisticus*', *Computers and the humanities* 14, 2 (October 1980), 83–90.
Calinescu, Matei, 'Orality in literacy. Some historical paradoxes of rereading', *Yale journal of criticism* 6, 2 (Fall 1993), 175–190.
Cohen, Daniel J., and Roy Rosenzweig, 'Web of lies? Historical knowledge on the internet', *First Monday* 10, 12 (December 2005).
de la Mare, Walter, 'Fare Well', in *Motley, and other poems* (London: Constable, 1918), 74–75.
Eccles, Mark, 'Middleton's birth and education', *Review of English studies* 7, 28 (October 1931), 431–441.
Evans, G. Blakemore, '"The shard-borne [-born] beetle", *Macbeth*, 3.2.42–43', *ANQ* 18, 4 (Fall 2005), 31–34.
Hawthorne, Nathaniel, *The scarlet letter. A romance* (Boston: Ticknor, Reed & Fields, 1850).
——, *The scarlet letter*, ed. Fredson Bowers (Columbus: Ohio State University Press, 1962).
Housman, A.E., *Selected prose*, ed. John Carter (Cambridge: Cambridge University Press, 1961).
Kane, George, 'Conjectural Emendation', in *Chaucer and Langland. Historical and textual approaches* (London: Athlone, 1989), 150–161.
Kermode, Frank, 'Divination', in *An appetite for poetry. Essays in literary interpretation* (London: Collins, 1989), 152–171.
Lass, Roger, *Historical linguistics and language change* (Cambridge: Cambridge University Press, 1997).
McGann, Jerome J., *A critique of modern textual criticism* (Chicago: University of Chicago Press, 1983).
Middleton, Thomas, *The collected works*, gen. eds Gary Taylor and John Lavagnino (Oxford: Clarendon Press, 2007).
Parker, Hershel, and Bruce Bebb, 'The CEAA. An interim assessment', *Papers of the Bibliographical Society of America* 68 (1974) 129–148.
Pizer, Donald, 'On the editing of modern American texts', *Bulletin of the New York Public Library* 75 (1971), 147–153.
Shakespeare, William, *Mr William Shakespeares comedies, histories, & tragedies* (London: Isaac Jaggard and Edward Blount, 1623).
Spencer, Matthew, Elizabeth A. Davidson, Adrian C. Barbrook, and Christopher J. Howe, 'Phylogenetics of artificial manuscripts', *Journal of theoretical biology* 227, 4 (21 April 2004), 503–511.
Tanselle, G. Thomas, Untitled review, *Common knowledge* 14, 3 (Fall 2008), 488.

Taylor, Gary, 'Inventing Shakespeare', *Deutsche Shakespeare-Gesellschaft West Jahrbuch* (1986), 26–44.

——, 'Praestat difficilior lectio. *All's well that ends well* and *Richard III*', *Renaissance studies* 2, 1 (March 1988), 27–46.

——, '*A game at chess*. General textual introduction', in *Thomas Middleton and early modern textual culture. A companion to the collected works*, gen. eds Gary Taylor and John Lavagnino (Oxford: Clarendon Press, 2007), 712–873.

THE REMARKABLE STRUGGLE OF TEXTUAL CRITICISM AND TEXT-GENEALOGY TO BECOME TRULY SCIENTIFIC

Ben Salemans

In the nineteenth century Karl Lachmann introduced a new text-critical method. The Method of Lachmann prescribes that the textual scholar must develop a stemma before starting any emendation activities. With the help of stemmas, this method has considerably reduced the number of unverifiable and unscientific judgements whether textual variants are original or corrupt. Nevertheless, the Method of Lachmann still had an unsystematic, unverifiable and therefore unscientific element: it uses so-called common errors for the development of a stemma. From about the 1980s mathematicians were able to build a chain, inductively, out of almost any set of textual variants. Textual scholars considered these chains to be, almost by definition, objective. It seemed that textual criticism had finally become a part of true science. The present contribution challenges the scientific character of the so-called objective, inductive approaches and presents the author's own neo-Lachmannian approach, which is influenced by biological cladistic insights about building genealogies. The theory it presents is subjective and deductive, but its subjectivity is fully accounted for, and the results can be verified.

Inductive and deductive research

Participants to the colloquium whose papers are included in the present volume were asked to reflect on a number of basic questions. Two of those questions were: 'is the innovation brought about by IT really a methodological innovation?' and, 'do we speed up classical techniques, or do we develop a new domain of techniques for access to the (classical) texts?'

Both questions deal with the philosophy of science and I need the philosophical notions of 'induction' and 'deduction' to answer them. Nowadays, induction, involving the application of statistical and/or mathematical methods to objective facts, is very popular. Deductive research, starting with the definition of subjective thoughts, ideas, hypotheses about the material or facts to be investigated, with the facts variously confirming, rejecting or requiring adjustment of the hypotheses, has a negative aura. That objectivity, and therefore induction, rules is a problem for many scholars in the humanities, who traditionally work with deductive methods. Because they do not want to be accused of subjective, and therefore unscientific, research, many researchers in the fields of the humanities feel obligated to replace their deductive research by inductive research. However, Karl Popper has clearly described how deductive research can be performed in a scientifically acceptable way.

Now, let us return to the questions cited above. As far as text-genealogical research is concerned, my answer to both questions is that IT offers us the new opportunity to rebuild old deductive, subjective and unverifiable—and therefore scientifically unacceptable—methods into modern deductive and scientifically fully acceptable methods. In short, my message is that deductive scholarly research can be made trustworthy through the use of the computer. This seems to me to be a methodological innovation. In other words, I am convinced that IT can cause a revival, or revolutionary continuation, of many traditional, perhaps old-fashioned, research methods in the humanities. That is, of course, good news for old-fashioned scholars in the humanities who are confronted with the problem that they are more or less obligated by their modern employers to turn their unscientific deductive research into inductive, empirical and mathematised, true scientific research.

Popper suggests that there are at least two conditions under which it is acceptable that a scholar uses subjective ideas or hypotheses as his or her starting point of research. The first condition is that the scholar formulates his or her hypotheses in a fully clear and controllable way. The second condition is that the application of these hypotheses is intersubjective, which means that the hypotheses can be applied and repeated by other scientists. In other words, trustworthy, scientifically acceptable, deductive research must be based on a fully clear theory (i.e. a well-formulated set of hypotheses), and its application must be fully controllable and repeatable. Once we have taught the computer to perform our hypotheses or deductive theory, both conditions are met. In the following I will try to demonstrate how this may be done. For that purpose, I will use parts of my own PhD research on the text-genealogical relationship of fourteen different text versions of the Dutch medieval play Lanseloet van Denemerken.[1]

The struggle of textual criticism to become 'scientific'

Textual criticism is the art of finding and/or removing unoriginal textual elements in texts. Such unoriginal (or: 'corrupt') textual elements are thought to have been introduced into the text by copyists or printers of previous versions of the text.

During the past centuries textual criticism has been performed in different ways. Until the nineteenth century, unoriginal elements in text versions were detected in a subjective, unverifiable, and therefore unscientific way. If a textual critic had the feeling (on the basis of his good taste, experience or *Fingerspitzengefühl*) that a certain part of a text was unoriginal or corrupted, he emended it. On the basis of this unverifiable, and therefore unscientific, feeling he corrected the part of a text that he presumed to be corrupt into what he believed to be the original reading intended by its author.

In the nineteenth century, the German philologist Karl Lachmann (and his successors, especially Paul Maas) introduced a new text-critical method, known as the Method of Lachmann. This method enabled scholars to determine in a more reliable and scientifically more acceptable way whether a part of a text was original or corrupt, without having to use unverifiable and unsystematic *Fingerspitzengefühle*. The Method of Lachmann prescribes that a text-critical editor first has to develop a stemma or text-genealogical tree, before he starts his emendation activities. The stemma expresses the historical relationship of the several extant versions of a text. It shows, for instance, that a certain text version belongs to the same 'family' of text versions which share a common ancestor (which may be lost). Often a stemma clearly dictates whether a part of a text in a text version is original or corrupted. With the help of stemmas, the Method of Lachmann

---

[1] For these details see my 2000 PhD dissertation, *Building stemmas*, which is available in pdf format at http://neder-l.nl/salemans/diss/salemans-diss-2000.pdf; appendices: http://neder-l.nl/salemans/diss/salemans-diss-2000-appendices.pdf.

considerably reduced the unverifiable and unscientific judgements by scholars about the originality of textual variants. This was a scientific improvement.

Nevertheless, the Method of Lachmann still had an unsystematic, unverifiable and therefore unscientific element: it uses so-called *common errors* in text versions for the development of a stemma. A common error is, roughly speaking, a textual element that shows up in two or more text versions and is corrupt/unoriginal, while all the other text versions have the original textual element. In other, modern, words: a common error is a defect in the DNA of a text version which is inherited by all descendants of that text version in which the DNA defect occurred first. Once you know enough DNA defects, or common errors, you know the genetic relationship of the text versions, yielding the stemma. The problem, the flaw in the Method of Lachmann was that the judgement on the originality of textual differences between text versions (or: 'variants') was still based on the unsystematic and unverifiable *Fingerspitzengefühl*.

In the twentieth century, several textual scholars, including W.W. Greg, Dom H. Quentin, V.A. Dearing, A. Dees and G.P. Zarri, conquered this Lachmannian flaw. They found that the base shape or deep structure of a stemma, the so-called *chain*, can be built without the unscientific judgement on the originality of textual variants. Indeed, a stemma can be built rather easily out of a chain. Again textual criticism seemed to have become more scientifically acceptable.

But then, from about the 1980s, something began to change. Modern textual scholars did not want to be accused of unscientific behaviour anymore. They wanted to act like and be respected as true scientists, like mathematicians. They met friendly colleagues from mathematical departments of their universities (true scientists) who were willing to build, inductively, *chains* (the deep structures or rough shape of stemmas) out of almost any package of textual variants ('facts'). Many modern textual scholars were thrilled: in their view, chains produced by mathematicians had to be, almost by definition, objective. At last all subjective flaws of earlier text-critical methods had been resolved. The ultimate goal had been achieved: at last, textual criticism (and the building of text-genealogical trees) had become a part of true objective science.

However, the question is whether inductive science, the art of teasing information out of gathered objective facts, is necessarily always scientific. In inductive research, the objective facts must be related to the goal of the research. If I want, for example, to predict the weather, I can gather all kinds of objective facts, such as paperclips, stones and bits of paper in and around my house. It is obvious that I will not be able to predict the weather with these facts, even though they are *objective*. In other words, a goal-oriented justification is necessary for the selection of objective facts in inductive research. I submit that so far a scientifically necessary justification for the use of all the variants as building blocks for historical trees has not been presented in inductive stemmatology. I think that in actual fact only very few variants can be used for the development of text-genealogical, historical, trees (see below). Some inductive, statistical stemmatologists admit that their trees are not historical trees, but trees which show the spread of the variants in the text versions. In that case, what is the virtue of such trees? Furthermore, statistical stemmatologists use all kinds of variants as elements to build their trees. Very often they remove variants out of their apparatus of facts, because they are believed to cause bias and complicate the development of their trees. Most of the time statistical stemmatologists consider this removal of difficult variants to be a part of the game. They hardly ever explain on what grounds they remove these variants. Therefore I generally consider these removal activities as unverifiable and subjective interventions. Inductive scientists have often accused deductive scientists of unscientific behaviour with their unverifiable subjective hypotheses. However, that same accusation applies here. Is inductive science in practice always synonymous with true objective science? I have my doubts.

Building a stemma of the Lanseloet van Denemerken *(1)*

Around 1990, after a three-year study of different text-critical methods, I thought I knew enough about building chains and stemmas. It was time to finish my PhD research and to develop the chain and the stemma of the fourteen text versions of the medieval Dutch drama play *Lanseloet van Denemerken.*

First, I developed Snobol/Spitbol software to produce a synoptic edition of the Lanseloet text versions. In this synoptic edition each verse of text version '2' (a Lanseloet text with 927 verses printed by Govert van Ghemen in Gouda around 1490) is presented together with the corresponding verses of the other thirteen text versions. I thought that with this synoptic edition it would be quite easy to detect the textual differences or variants between the fourteen text versions.

Secondly, I started to write down, by hand, the textual differences between the fourteen text versions. At that time, I thought it would take me a few weeks or months to find the significant variants in the Lanseloet corpus. Once I had found them, I could start to build the chain and the stemma of the fourteen Lanseloet text versions. In those happy days of ignorance I really felt able to isolate these significant variants. And I actually did find significant variants rather easily. The unimportant variants were easily detected as well.

Just when I thought that I could finish my dissertation quickly, I was confronted with a double crisis in my research that took a further ten years to resolve. The first crisis was that I was confronted with the fact that I did not have a clear system to reject or accept variants which I could use to build the Lanseloet chain and stemma. For instance, when I arrived at the twentieth verse of the synoptic text edition, I decided to use a certain variant for the development of the Lanseloet chain. But while accepting that variant, I realised that I had rejected a comparable variant in the previous nineteen verses. Therefore, I had to restart my search for significant variants all over again. It took me about twenty restarts to accept the fact that I needed a clear system, a theory, by which 'good' and 'bad' variants could be distinguished on a principled basis.

The second crisis came simultaneously. It was the nagging question why a certain variant was 'good' or 'bad'. It really took me a few months to accept the fact that I actually did not understand why a variant between the text versions could provide information about the historical relationship of these text versions. In other words, I was confronted with the fact that I did not know precisely enough what a 'text-genealogical variant' or 'text-historical informative variant' was. I was gathering all kinds of variants, but what were they supposed to prove?

Of course I could have chosen to go with my temporary set of variants to the inductive mathematician Evert Wattel at the Free University of Amsterdam. Wattel had developed statistical software that could draw beautiful chains out of almost any set of variants. I chose not to do so. What would that Lanseloet chain be worth? How could such a chain (or stemma) be a historical tree in which historical relationships between the text versions were expressed if I was not sure that the building elements of such a tree, the variants, were revealing significant relationships?

New insights from cladistics

It was in those days that I met a biologist from the University of Amsterdam, Dr Willem Ellis. He had read some of my articles on building text-historical trees, and he saw similarities between my text-genealogical research and his biological taxonomy, a part of biology which produces family trees of animals and plants. Dr Ellis gave me a lot of publications on biological taxonomy: a whole new world, with new insights opened before me.

I felt attracted to a rather new biological taxonomic movement called 'cladistics', first developed by the German professor Willy Hennig. The core of cladistics is the permanent question which elements or characteristics in a species can be used to develop genealogies, the equivalent of chains and stemmas in text criticism. The simple lesson taught by cladistics is that we must be very careful in using characteristics for genealogical research. For example, the fact that both swallows and flies have wings, does not imply that these birds and insects belong to the same family. The characteristic 'having wings' is not a trustworthy genealogical informant. Cladists stress that finding causally linked characteristics is a very difficult and delicate task. From cladistics I learned that my doubts about using certain variants to build text-historical trees had been justified. Furthermore, I decided, based on cladistic insights, that I would only use variants as building elements for my Lanseloet chain if I was sure that these variants have relationship-revealing powers. This would imply that most of the Lanseloet variants could not be used for the development of the Lanseloet tree.

After my excursion into cladistics I suddenly observed a lot of doubtful variants and only very few useful variants in my computer-produced synoptic edition of *Lanseloet van Denemerken*. Apparently I did have ideas about how informative and non-informative variants could be recognised, based on my reading of numerous text-genealogical studies. Unfortunately, I did not know of any study in which all characteristics of genealogical and ungenealogical variants were described in detail, so I decided to develop such a list of characteristics.

I also decided to identify all these characteristics by number; this enabled me to justify why I accepted or rejected a variant as a building
element for the Lanseloet chain. In the synoptic edition I placed a plus
sign (meaning accepted) or minus sign (meaning not accepted) after
the number of the relevant characteristic: for instance, '6a-' meant 'this
variant is rejected because of characteristic 6a'.

My list of genealogically useful and non-useful characteristics contains eleven characteristics (or twenty, if subcategories are counted
separately) They are presented in the following table:

Table 1: Text-genealogically useful and unuseful characteristics[2]

Characteristic 1: Text-genealogical variants belong to the same variation
place.
Text-genealogical variants are textual differences, in preferably single words
that occur in the *same variation place*.
Characteristic 2: Text-genealogical variants are part of type-2 variations.
Characteristic 3: Text-genealogical variants stand in a grammatically adequate
environment.
Characteristic 4 (4a, 4b): Text-genealogical variants belong to the same word
categories of substantives or verbs.
  4a. Text-genealogical variants belong to the *same word classes*;
  4b. They are *substantives* (= *substantive nouns*) or *verbs*, except auxiliaries
  (in Dutch: 'hebben', 'zijn', 'zullen', 'willen', etc.).
Characteristic 5: Text-genealogical variants can belong to other word categories when standing in rhyming position in verses.
Characteristic 6 (6a, 6b, 6c): Text-genealogical variants are not accidentals or
small spelling differences.
The textual differences between text-genealogical variants can *never* be *accidentals*, like:
  6a. Orthographical or diacritical differences;
  6b. Differences in word boundaries are considered to be orthographical,
  and thus accidental;
  6c. Nonsense words, obvious slips of the pen of the typesetter, or clearly
  incorrect, corrupt words (or word sequences that are semantically
  obviously incorrect) that can be changed quite easily into correct
  words (or word sequences).
Characteristic 7 (7a, 7b, 7c, 7d): Text-genealogical variants are not potential
regional, ideolectic, diachronic or other parallelisms.
Textual differences are not genealogical when it is possible or probable that
these differences are parallelistic. Apart from the accidental (orthographical) parallelism, mentioned here as characteristic 6, we recognize four

---

[2] Table adapted from Salemans, *Building stemmas*, 111–12.

Table 1 (*cont.*)

other, not strictly monolithic, *types of parallelism*: inflectional parallelism ('is'/'was'); synonymous and ideolectic parallelism ('white'/'pale'); regional parallelism ('color'/'colour'); diachronic or historical parallelism. To minimize the danger of parallelism, we formulate the following rules:

7a. The differences between genealogical variants cannot be mere differences in inflection;

7b. The difference between variants or their kernels/roots must not merely be a (phonetic) difference in a *range of vowels*;

7c. The differences between variants cannot be mere differences between the different vocabularies of languages or dialects;

7d. The (cores of the) variants must be *rare*, in the sense that it is not likely that a variant can be easily interchanged with another variant. This rule attempts to prevent the use of synonymous parallelisms, like for instance variants of the names of well-known people or things. Some *common* words can be turned into other more or less synonymous words easily, especially words that express an assertive act, like 'say', 'speak', 'tell', 'il fait', 'il dit'.

Characteristic 8: Text-genealogical variations in word order.

A difference in a syntactically adequate *word order* must be considered to be a genealogical variation, as long as the difference in word order does not merely concern a different placement of an adverb in a verse or sentence.

Characteristic 9 (9a, 9b): In verses, rhyming conventions must be obeyed.

9a. When text-genealogical variants are part of rhyming texts and are in rhyming position, they have to *obey (at least assonant) rhyming conventions*. (Source: first rule, element 'fits well and inconspicuously'.) If they violate them, this may be due to the interpolation/omission of one or more verses. Such a deletion/addition can be genealogically important;

9b. A special case of the violation of rhyming conventions occurs when one verse ends with a certain rhyming word and the immediately following verse ends with the same word. The philologist must study these verses with *duplicate rhyming words* closely, because it is very likely that these verses or the surrounding verses contain an error which occurred during the transmission process.

Characteristic 10: Inversion of verses.

The *inversion* of (the rhyming words in) verses is genealogically informative when these verses fit well in the text.

Characteristic 11 (11a, 11b): Addition and omission of words and verses.

11a. The *addition* (or *interpolation*) and omission of *words* is genealogically informative when these words fit well or offer no crucial information. Notice that the presence or absence of small frequently used words (like 'so') does not give text-genealogical information;

11b. The *addition* (or *interpolation*) and omission of complete *verses* is genealogically informative when these verses fit well or offer no crucial information.

These eleven, or twenty, characteristics did not just spontaneously appear. My dissertation explains how I found them. I consider these characteristics to be my text-genealogical theory, which describes how to evaluate the text-genealogical value of a variant. I must stress that these characteristics are hypotheses that must be applied and tested in a Popperian way: they can alternatively be rejected, falsified, or refined.

Building a stemma of the Lanseloet van Denemerken (2)

With my new list of characteristics I restarted my search for text-genealogical variants in the synoptic edition of the fourteen Lanseloet text versions. Above every variant I noted in pencil the number of the characteristic and a minus or plus. Indeed my characteristics or text-genealogical hypotheses were subjective, but now they could be used and checked by other scholars as well. The characteristics were intersubjective claims and therefore they were scientifically valid. But then I was confronted with a new problem. There were so many variants in my synoptic edition that it was almost impossible to present them in my dissertation clearly. To deal with this problem I decided to develop software (again written in Spitbol/Snobol) that would be able to recognize the variants and judge the variants according to the number of relevant characteristics.

The development of this software took me some years, but the end-result was worthwhile: I managed to let the computer perform my theory. Thus, my subjective deductive theory had become systematic and therefore repeatable, meeting another key criterion in Popper's account of scientific research.

Figure 1 gives an example of how the computer software treated the variants in the synoptic edition near verse 02.168 (verse 168 of base text 2 printed by Govert van Ghemen). At the top of the picture, we see, on the right, the synoptic text with verse 02.168 and the corresponding verses of the other thirteen Lanseloet texts. Notice that the fragmented texts 3 and 4 do not have any text here.

To the left of the synoptic text we see the shorthand version of the synoptic verses. In the Middle Ages and in the sixteenth and seventeenth century there was not one overall accepted orthography, which prescribed how words had to be spelled. The computer uses the shorthand version of the versions to find variants, not hindered by small spelling differences.

```
Shorthand version of synoptic text          Synoptic text
-----------------------------------------   -----------------------------------------
02.168 ende si is van live so grasios      < 02.168 Ende si is van liue soe gracioes [149]
01.211 ende van live so gratiose           < 01.211 En(de) van live soe gratioyse [190]
05.169 ende si es van live so gratios      < 05.169 Ende si es van liue so gratioes [fo.-A4v-]
06.170 int si is van live so grasios       < 06.170 Ind sy is van liue so gracioes
07.170 int is van live so grasios          < 07.170 Ind is van liue soe gracioes
08.170 int is van live so gratios          < 08.170 Ind is van lijue so gracioes
09.167 ok si is van likham so gratius      < 09.167 Ooc zy is van Lichaem soo Gratieus
10.174 ok si is van likham so gratius      < 10.174 Oock sy is van Lichaem soo gratieus/
11.176 ok is si van likham so gratius      < 11.176 Oock is zy van Lichaem soo gratieus
12.175 ok is si van likham so gratius      < 12.175 Oock is sy van Lichaem soo gratieus/
13.175 ok is si van likham so gratius      < 13.175 Oock is sy van Lichaem soo gratieus/
03*                                        < 03=>*
04*                                        < 04=>*
14.154 ok is si van likham so gratius      < 14.154 oock is sij van lichaem soo gratieus

Simple observations:
---------------------
02.0168 ob01: 02-06-09-10-|11-12-13-14-: W.O "si"-"is" (T2);8;
02.0168 ob02: 01-02-05-: "ende"
02.0168 ob03: 02-05-06-09-10-11-12-13-14-: "si"
02.0168 ob04: 02-06-07-08-09-10-11-12-13-14-: "is"
02.0168 ob05: 01-02-05-06-07-08-: "live"
02.0168 ob06: 02-06-07-: "-grasios"
02.0168 ob07: 05-08-: "-gratios"
02.0168 ob08: 06-07-08-: "int"
02.0168 ob09: 09-10-11-12-13-14-: "ok"
02.0168 ob10: 09-10-11-12-13-14-: "likham"
02.0168 ob11: 09-10-11-12-13-14-: "-gratius"

Formulas:
---------
    1.  02.0168 08 ?obs01: 02-06-09-10-|11-12-13-14-: W.O "si"-"is" (T2);8;
    2.  02.0168 06 ^comb.: 01-02-05-|06-07-08-="ende"|"int" (^:W1=Co;W2=Co;4a+;4b-)
    3.  02.0168 09 ^comb.: 01-02-05-|09-10-11-12-13-14-="ok" (^:W1=Co;W2=Av;4a-)
(   a.  02.0168 09 ^comb.: 01-02-05-|09-10-11-12-13-14-="ende"|"likham" (^:vp;1)
(   b.  02.0168 09 ^comb.: 01-02-05-|09-10-11-12-13-14-="ende"|"-gratius" (^:vp;1) )
(   c.  02.0168 12 ^comb.: 01-02-05-06-07-08-|09-10-11-12-13-14-="live"|"ok" (^:vp;1) )
(   d.  02.0168 12 ^comb.: 01-02-05-06-07-08-|09-10-11-12-13-14-="live"|"likham" )
    4.  02.0168 12   comb.: 01-02-05-06-07-08-|09-10-11-12-13-14-="live"|"-gratius" (^:vp;1)
    5.  02.0168 05   comb.: 05-08-|02-06-07-="-gratios"|"-grasios" (wds r.p;5)
(   e.  02.0168 09 ^comb.: 02-06-07-|09-10-11-12-13-14-="-grasios"|"ok" (^:vp;1) )
(   f.  02.0168 09   comb.: 02-06-07-|09-10-11-12-13-14-="-grasios"|"likham" (^:vp;1) )
    6.  02.0168 09   comb.: 02-06-07-|09-10-11-12-13-14-="-grasios"|"-gratius" (wds r.p;5)
(   g.  02.0168 08 ^comb.: 05-08-|09-10-11-12-13-14-="-gratios"|"ok" (^:vp;1) )
(   h.  02.0168 08 ^comb.: 05-08-|09-10-11-12-13-14-="-gratios"|"likham" (^:vp;1) )
    7.  02.0168 08 ?comb.: 05-08-|09-10-11-12-13-14-="-gratios"|"-gratius" (?<:vow.;7b)
    8.  02.0168 09 ^comb.: 06-07-08-|09-10-11-12-13-14-="int"|"ok" (^:W1=Co;W2=Av;4a-)
(   i.  02.0168 09 ^comb.: 06-07-08-|09-10-11-12-13-14-="int"|"likham" (^:vp;1) )
(   j.  02.0168 09 ^comb.: 06-07-08-|09-10-11-12-13-14-="int"|"-gratius" (^:vp;1) )
    9.  02.0168 12 ^treat: 2ch.wrd "si" (Pn) in 02-05-06-09-10-11-12-13-14- (rest: 01-07-08-);
        (+)                (^:small word?);11a?
   10.  02.0168 12 ^treat: 2ch.wrd "is" (Au) in 02-06-07-08-09-10-11-12-13-14- (rest: 01-05-) (T3
        (+)                or T2?); (^:small word?);11a?
   11.  02.0168 12 ^treat: 2ch.wrd "ok" (Av) in 09-10-11-12-13-14- (rest: 01-02-05-06-07-08-);
        (+)                (^:small word?);11a?
```

Figure 1: Treatment by the software of (synoptic) Lanseloet verses 02.0168.[3]

Below the horizontal line, under the synoptic and shorthand text, the computer gives simple observations: the computer detected that there is a difference in the order of the words 'si' and 'is' in the verses. In texts 2 (verse 02.168), 6 (verse 06.170), 9 (verse 09.167) 'si' comes before 'is', while in the verses of texts 11, 12, 13 and 14 'si' comes after 'is'. Characteristic 8 deals with differences in word order, claiming that this is an important, text-genealogical variant. Look to the right side of

---

[3] Figure from Salemans, *Building stemmas*, 122–23.

the rule in 'ob01' (for 'observation 01'), where ';8' denotes that characteristic 8 is applied.

Then observations 2 to 11 ('ob02'...'ob11') follow. These observations concern the variants in the Lanseloet texts which occur in at least two text versions. For reasons which I will not discuss here, I am not very interested in variants that occur in only one text version. Observation 2 (ob02) tells us that texts 1, 2 and 5 have the variant 'ende' in common.

Under the second horizontal rule we see 'formulas'. The first formula concerns the difference in word order, which we have already discussed. Note the question mark before 'obs01'. This question mark indicates that I have doubts about the correctness of word order characteristic 8. Because of these doubts I will not use characteristic 8, concerning word order variants, as relationship revealing, and text-genealogically informative. Once a trustworthy chain of the Lanseloet text versions has been built, we can check whether characteristic 8 is valid or should be revised or rejected.

The second formula gives us the following information: texts 1, 2 and 5 have 'ende', while texts 6, 7 and 8 (these three texts were printed in Germany, Cologne) have 'int'. Then the software starts to investigate whether these variants can be used as variants to build the Lanseloet tree. Of course the software uses the eleven (or twenty) characteristics for this determination. We read in the output '(^: W1=Co;W2=Co;4a+;4b-)'. This must be understood as follows: W1, the first word ('ende'), is a conjunction ('Co') and W2, the second word ('int'), is a conjunction ('Co') as well. Characteristic 4a says that genealogical variants must belong to the same word class; 'ende' and 'ind' are both conjunctions and therefore characteristic 4a is met; therefore, the test with characteristic 4a is valid, which is denoted as '4a+'. Then the computer looks further and tests whether the variants agree with characteristic 4b, which roughly says that variants must be nouns or verbs. Because both variants are conjunctions characteristic 4b is negative: '4b-'. Therefore both variants are rejected, expressed by the '^' sign. Incidentally, by mentioning '4a+' even though it is overruled by '4b-' the computer enables me to evaluate characteristic 4a later on.

The computer tested many thousands of variants. The test resulted in about two hundred variants which were in agreement with the characteristics of useful text-genealogical variants. Due to some small software errors and due to the fact that three text versions were in a German dialect, I had to discard by hand about one hundred and fifty variants. With the final 54 variants the computer could build a Lanseloet chain without any bias or contradiction: all 54 variants pointed at one and the same chain. Though bias is counteracted by the computational technique this is not absolute proof that the chain is correct. But at this moment there is reason to assume neither that the chain is incorrect nor that my thoughts about text-genealogical characteristics are incorrect.

Finally, I evaluated all characteristics with the help of the computer. As I mentioned before, some characteristics prohibit the use of certain variants as chain building elements. The evaluation taught me that these prohibiting characteristics did work well: they prevented the use of variants which are clearly in contradiction with the Lanseloet chain. I also mentioned that I had doubts about characteristic 8 on word order. The evaluation shows that some word order variants are in clear contradiction with the Lanseloet chain. Therefore, we conclude that word order variants must not be used as chain building elements. Perhaps we should study under what special circumstances word order variants may be used for chain development. Of course my theory is not finished and it must be tested agaٕints other texts, so that it can be improved.

Conclusion

At the end of this contribution, I will return once more to the two questions quoted at the beginning of this chapter. I hope that I have demonstrated that IT enables us to turn unscientific deductive theories into fully verifiable, intersubjective and repeatable, and therefore scientifically acceptable, deductive theories. This is, of course, an important improvement. Possibly, this innovative IT approach may open new domains in the traditionally deductive fields of the humanities. Perhaps some other deductive and unscientific theories in the humanities can also be offered new scientific lives with the help of the computer.

Reference

Salemans, Ben, *Building stemmas with the computer in a cladistic, neo-Lachmannian, way. The case of fourteen text versions of Lanseloet van Denemerken: Een wetenschappelijke proeve op het gebied van de Letteren* (Nijmegen: the author, 2000), 111–12.

PART THREE

CASE STUDIES

SEEING THE INVISIBLE:
COMPUTER SCIENCE FOR CODICOLOGY

Roger Boyle and Hazem Hiary

Computer science exhibits very rapid development, and when coupled with interdisciplinarity, it can bring considerable benefit to other fields of study. We present a particular example in watermark location and identification. Computation in this area of work is not new, but we show that digitally native acquisition and model-based inspection permit new results to be drawn from documents examined by hand some decades ago. Some ideas we present are generalisable to other aspects of the archaeology of paper manufacture, while others extend to humanities research more broadly. Our conclusion is that the merits of dialogue between scholars of the humanities and computer scientists will grow significantly over time. Such dialogue between communities is often impeded by an absence of common language, and the real problem may be the development of suitable hybrid scholars.

Introduction

Since its inception, computer science has developed exceptionally fast as a discipline. This is true not just in its technical depth, but also in its broad potential. At superficial levels consumer products in the high street exhibit the leaps made by computer science research, but academies worldwide have over the last ten years also fostered ever more enthusiastically forms of interdisciplinary working that were facilitated by the spread of computational methods.[1] This is not new, but the strategic encouragement and considerable resourcing behind these new levels of collaboration are a recent phenomenon. Aside from long-established and obvious links with traditional engineering and science subjects, computer science is now very evident in geography, medicine, linguistics and other academic fields.

---

[1] There is a rash of recent literature studying this phenomenon; just one example is A. Repko, *Interdisciplinary research.*

At the forefront of this drive lie well-resourced activities: hardware and software laboratories can be expensive facilities to run, and the staff within them can command exceptional salaries. The comparatively lower levels of remuneration available for computational innovation in the arts and humanities may help explain the lesser evidence of partnership of these areas with computer scientists. Paradoxically, consumer market developments can be a salvation for resourcing arts and humanities disciplines—computers are now cheap and usable. As time goes by they become even cheaper, while their esoteric and redundant features become less of an obstacle for their use. Software packages of extraordinary power can be obtained at modest or no cost, and everyone can join in the fun. At the same time, computer science is delivering phenomenal and cheap storage capacity and the wherewithal to move data around the planet with imperceptible delay. Digital repositories of startling size are springing up in the library world, while scholars with a standard desktop machine can obtain exceptionally fast Internet access internationally to materials that hitherto have required international travel and physical inspection.

It is thus no surprise that arts and humatities scholars are making extensive use of computers. The issue may be that much of what is being done with consumer products could be amenable to the products of recent research to deliver better and more robust solutions. And in the true spirit of interdisciplinarity, the spread of computer science innovations into new domains can frequently generate new and productive research for the computer scientists as well. We note then a risk of lost opportunity: if cross-fertilisation between computer science and arts and humanities computing is not encouraged, academic areas may be deprived of fruitful approaches to their problems. Promising lines of research inquiry may then become neglected or abandoned altogether for lack of innovation in the area of computation.

In this paper, we offer comment on the pace of arts and humanities research innovation that accompanies straightforward technical development, and consider how this may inform under-explored areas such as codicology and palaeography. In particular we consider watermarking, which has long been of interest to codicologists. Using a challenging dataset we illustrate how a computer science approach can provide a robust solution to a difficult problem, and reveal paper details hitherto unseen. Our conclusion is that ongoing dialogue between humanities scholars and computer scientists is to the benefit of all and deserves nurturing.

Computer science and artificial intelligence; their role in other domains

Teresa Nakra wrote that 'The Holy Grail of computer science is to capture the messy complexity of the natural world and express it algorithmically', confirming that there exist many areas in which properly conducted computer science might contribute significantly outside its own boundaries, while at the same time permitting development of its core activity in formality and algorithms.[2] Unsurprisingly, such developments are most evident in the natural sciences and engineering: excellent recent examples are visible in medicine with the analysis of immense and difficult data derived from modern imaging modalities[3] and in biology, where evolutionary systems[4] and more recently protein analysis[5] have yielded significant insights to computational techniques.

In the less quantitative disciplines, it is often artificial intelligence which has been seen to contribute most successfully; for example, major advances are being seen in automated CCTV monitoring and the understanding of natural language,[6] where sophisticated techniques that even the intelligent lay user would have difficulty in developing[7] are beginning to see routine use. There is no doubt that some early attempts at complex tasks such as machine translation were laughably inadequate, but the 2010 state of the art is no longer so, leaving challenges such as automated reading of scripts such as Arabic an attainable ambition.

Pragmatics dictate that productive application of modern computer science techniques to the humanities has been slow: most seriously, the research funding is lower and scarcer. Nevertheless, humanities computing has in recent years developed significantly: a good early indicator has been the establishment of institutes such as the Centre for Computing in the Humanities at King's College London[8] to focus and nurture expertise. More recently the definition in the UK of a

---

[2] Davidson, 'Measure for measure', 66.

[3] Various authoritative publications passim: 'IEEE transactions on medical imaging' and 'Current medical imaging reviews' (Bentham) are perhaps foremost.

[4] Bentley and Corne, *Creative evolutionary systems.*

[5] Bradford et al., 'Insights into protein analysis'.

[6] Dee and Velastin, 'How close are we?'.

[7] Sahami et al., 'A Bayesian approach'.

[8] CCH, Centre for Computing.

computer science 'Grand Challenge in research' *Bringing the Past to Life for the Citizen*[9] signals a level of academic communication that opens new channels and possibilities. Such initiatives are timely, since many new opportunities are opened. The combination of very cheap bulk storage and high bandwidth communication across the Internet make phenomenal quantities of information easily available on a global basis. Immense digital repositories and digitisations of very scarce material (Insciptifact[10] and Codex Sinaiticus[11] are two examples) are becoming more common.

More interesting, theoretical and algorithmic advances in computer science and artificial intelligence continue to develop at speed. Some of these new techniques are probably unexpected and consequently unconsidered or under considered in the humanities. While the immediate application of these major developments to digital text analysis is not always clear, experience teaches us that unexpected connections often exist.

Two examples will give substance to these points. Firstly, Terras has contributed significantly to the reading of ink and stylus texts discovered at the British Roman fort of Vindolanda (see Figure 1).[12] The work is ground-breaking in aiding historians to read ancient texts, but it was conducted as computer science research, deploying a state of the art artificial intelligence architecture that exploited information elicited from domain experts. And secondly, Huang et al. demonstrate well the potential for modern computational techniques in difficult papyrological problems.[13] Nineteenth-century governmental ledgers in the National Archives of Singapore suffer from serious legibility problems as a result of ink bleed from verso to recto. Elementary approaches to increasing their legibility might involve simple algorithms that are easily accessible via Photoshop and similar software, but these attempts have limited scope given the variability of text intensity and subtlety of the problem. The authors show how the problem is amenable to attack through the

---

[9] UKCRC, Current Grand Challenges.

[10] Insciptifact, Image database. See also the chapter 'Concrete abstractions' by Leta Hunt, Marilyn Lundberg and Bruce Zuckerman in this book.

[11] Codex Sinaiticus. See also the chapter on the Codex Sinaiticus by David Parker in this book.

[12] Terras, 'Interpreting the Image'; Vindolanda tablets online.

[13] Huang et al., 'Framework'.

Figure 1: An example tablet from the Vindolanda excavations. Reading of these very challenging texts was significantly aided by the application of AI techniques (Terras, 'Interpreting the image'). Reproduced with permission of Centre for the Study of Ancient Documents, University of Oxford.

Figure 2: Removal of inkbleed in Singaporean governmental ledgers. A coupled Markov Random Field approach has removed the verso interference where simple thresholding would have failed. Reproduced with permission.

application of current algorithms popular in the artificial intelligence pattern classification community (see Figure 2).

More specifically, the characterisation of *shape* is a key requirement in many visual tasks: computer recognition of shape is a difficult challenge. Recent innovative paradigms[14] for addressing the problem of computational recognition of shape that have developed in computer science have not to date been applied to comparable tasks in the humanities, such as watermark retrieval from online databases. Watermark retrieval is a task that preoccupies many codicologists in particular.

The study of watermarks in paper

Paper is schizophrenic. While its raison d'être is to carry inscriptions, paper has a character and a story of its own. The messenger can tell stories beyond the message it carries, so that there can be understanding without reading the text. This paper understanding is knowledge about 'the commodity that links all the individual warehouses, bookstalls, and printing houses.'[15] Gants explains the value of studying the material base of text documents: knowledge of paper manufacture, sale, transport and purchase has long assisted attribution and dating. This is achieved in particular by scrutinising the material of the paper carrier: its telltale signature of manufacture in laid and chain lines, and its watermark (see Figure 3). Coupled with archive data from paper-manufacturing mills it becomes possible to reconstruct the wholesale and retail routes that papers have taken, and by following these detailed routes it often becomes possible approximately to date documents.

In some cases this may be done with greater precision: in cases where the moulds used for producing paper sheets became damaged in use, the papers that were produced include highly specific internal patterns that date their manufacture more accurately. Such approaches require significant detective work in data collection to track evolving changes, and can generate problems for scholars when archived watermark examples differ (due to progressive damage) from those seen in specimens.[16] The difficulty is of course that such internal features are not easily visible, because their pattern distributes across individual pages and is affected by wear and tear and other factors. The work we

---

[14] Lowe, 'Object recognition'.
[15] Gants, 'Identifying'.
[16] Sosower, 'Greek manuscripts'; Sosower, *Signa officinarum*.

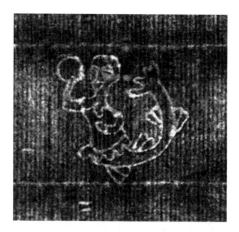

Figure 3: Features of paper—laid lines (vertical) and chain lines (horizontal) define the mould, inside which the watermark is effected by wire design. This example is taken from *Watermarks in Incunabula printed in the Low Countries*, http://watermark.kb.nl/.

describe in this paper is a study of watermark location and characterisation, but it is worth noting that the less glamorous laid and chain lines have been argued to be of greater value in characterising paper moulds.[17]

Paper watermarks, which are changes in paper thickness, have been in use for over seven hundred years, with the oldest known watermarked paper produced in 1282 in Fabriano.[18] They have been used as trademarks of paper-makers, as identification marks for sizes of moulds used in manufacture, as symbols of religious groups called Albigenses, as an aid to illiterate workmen, and as an exercise in imagination by paper-makers keen to show their artistic skills. Their use spread and watermarks came to be used to trademark paper, a proof of the date of manufacture and an indication of paper size, culminating in their use as a mark against counterfeiting on money and other documents.[19]

Two main types of paper watermarks exist: *line* (typically known as wire), and *shadow* (light and shade). *Combined* watermarks have both, and other varieties exist.[20] Wire watermarks are made using lines to form various patterns such as letters, numbers, portraits; they appear

---

[17] Gants, 'Identifying'; Stevenson, 'Chain-indentations'; Vander Meulen, 'Identification of paper'.
[18] Briquet, 'Notice sur le recueil'; Briquet, *Les filigranes*; Spector, *Essays*.
[19] Hunter, *Papermaking*.
[20] Loeber, *Paper mould and mouldmaker*; IPH, International Standard.

lighter than the surrounding paper area. Light and shade watermarks have patterns resulting from relief sculptures on the mould; these designs give the watermark further variations to support more features, and may appear lighter or darker than the surrounding area. In many mills, paper making was often accelerated by making pairs of moulds with two very similar but not identical watermark designs, causing them to feature as so-called twins.[21]

Pavelka writes, 'The study of watermarks is a seductive if somewhat esoteric pastime. While it is normally the beauty and aesthetic quality of watermarks that initially attract the researcher, they are more than just pretty affectations and can shed light on historic trends and events.'[22] Watermarks in paper have attracted a wide range of interest from researchers for centuries.[23] However, watermark designs are available not only in several different forms, but also dynamically change over time. This has introduced some complications that have hindered more systematic study of the artefacts. They only become visible to the eye when faced against light, and are usually obstructed by writing ink and other noise in paper.[24] Many approaches have been developed in order to reproduce and exploit them; Hiary provides a good survey of the ways in which instruments facilitate research on watermarks.[25]

Image collection techniques fall under four broad headings:

- *Manual*: The most primitive techniques would either place a document on a light table and a user would copy the watermark onto tracing paper laid on top, or alternatively a clean sheet is placed over the document and a pencil is rubbed over it with long diagonal strokes. These are simple techniques, but potentially damaging.[26]
- *Back-lighting*: This requires a high resolution, usually digital camera and a light source. The camera captures reflected (with normal light) and transmitted (with back-light from slim light or light box)

---

[21] Spector, *Essays*; Stevenson, 'Watermarks are twins'.

[22] Pavelka, 'Review'.

[23] Bower, *Turner's papers*; Gravell, *Catalogue*; Rouse, untitled review of *Missale speciale*; Spector, *Essays*; Stevenson, 'Shakespearian dated watermarks'; Stevenson, 'Paper as bibliographical evidence'; Stevenson, *Problem of the Missale speciale*.

[24] 'Noise' in images is something other than that which is sought; its source may be the imaged material—where it may or may not be the result of damage or interference—or part of the imaging process. Often, one person's noise is another person's signal.

[25] Hiary, Digital extraction.

[26] Ash, 'Recording watermarks'; Briquet, *Les filigranes*; Churchill, *Watermarks in paper*; Haupt, 'Wasserzeichenwiedergabe'; Heawood, *Watermarks*.

images of the paper.[27] It is quick and relatively low cost, and produces good, natively digital image quality without darkroom conditions. However, it captures all the details of paper, including the watermark and any other designs that may interfere with it.

- *Radiographic techniques*: Radiography offers the ability to display changes of paper thickness no matter what is printed on it.[28] Various techniques (beta-radiography, soft X-radiography, electron-radiography) are in use, but they are normally some combination of expensive, time-consuming, or potentially hazardous.[29]
- *Special purpose techniques*: A range of other, specialist approaches is or has been in use. These include: the Dylux method,[30] the Ilkley method[31] and thermography.[32] These techniques suffer from one or more of a number of drawbacks such as poor images, hazard to users or materials, and susceptibility to writing. They are rarely natively digital.

A number of web archives of watermarks exists, and to date most work on watermark extraction has been in pursuit of the compilation of such.[33] Manual techniques represent an end in themselves, but the more sophisticated approaches (and digitisations of manual collections) have been subject to further computer-based manipulation. Mainstream image processing can play a useful role: commonly, combinations of edge detection, region extraction and morphology are used to try to isolate clean watermark representations.[34] A few approaches attempt to build sophisticated models such as reflectance,[35] or seek

---

[27] Ash, 'Recording watermarks'; Bernstein, Memory; Christie-Miller, Early paper project; Stewart et al., 'Techniques'; Van Staalduinen, 'Comparing'; Wenger and Karnaukhov, 'Distributed database and processing system'.

[28] Van Aken, 'An improvement'; van Staalduinen et al., 'Comparing'.

[29] Van Staalduinen et al., 'Comparing'; Ash, 'Recording'; Van Hugten, 'Weichstrahlradiographie'; De la Passardière and Bustarret, 'Profil'; Riley and Eakins, 'Content-Based Retrieval'; Schnitger and Mundry, 'Elektronenradiographie. Ein Hifsmittel'; Schnitger et al., 'Elektronenradiographie als Hilfsmittel'; Schoonover, 'Techniques'; Small, 'Phosphorescence'; Tomimasu et al., 'Comparison'; Bridgman et al., 'Radiography of panel paintings'; Bridgman, 'Radiography of paper'; Ziesche and Schnitger, 'Elektronenradiographische Untersuchungen'.

[30] Gants, 'Pictures for the page'; Gravell, 'Wizard'.

[31] Schoonover, 'Techniques'.

[32] Meinlschmidt, 'Original or fake?'.

[33] Bernstein, Memory of papers; DUIAH, International database; Koninklijke Bibliotheek, Watermarks in Incunabula; WIES; WZMA.

[34] Sonka et al., *Image processing*.

[35] Stewart et al., 'Techniques for digital image capture'.

identifiable properties of paper such as (regular) chain lines when Fourier techniques—a mathematical technique that allows manipulation of regular patterns—are of use.[36] A primary aim is to match extracted watermarks to existing databases.[37] Most work however has experienced problems with noise in images, attributable to paper quality and interference from recto and verso inscription. Self-evidently, this will hamper work on many of the most interesting artefacts to be found in libraries. There is frequently a reliance on certain hand-picked parameter settings in a range of algorithms, resulting in a lack of systematicity that hinders broad applicability.

A case study: The 'Mahdiyya' Quran

In what follows we discuss a case study in which we tried to further the detection and inspection of watermarks. Our work addressed three specific problems: developing a technique to work on materials seen to date as very challenging; extracting complete, or near-complete, watermarks from collections of documents in which only fragments are easily visible or accessible; and developing tools that may be useful in distinguishing subtle differences, in particular twins.[38]

We tried to work in a context of parameter selection that is automatic or adaptive. We have chosen to use back-lighting because it is simple, relatively quick, digital and cheap. The data used has been selected from the Arabic holdings of the University of Leeds, which include a number of rare, unusual and little-known texts which all carry wire watermarks. The primary document is the Mahdiyya Quran.[39] This was written in 1881 in Sudan and exhibits thick writing strokes on very thick, uneven paper. The paper was cut before use, so that no page contains a complete watermark or countermark. Yet we know that it bears a double-headed eagle watermark of the Austro-Hungarian Empire with the countermark 'Andrea Galvani Pordenone'. The watermark has a shield containing a moonface. Plate 10 illustrates an example page.

---

[36] Gants, 'Identifying'; Toft, The Radon Transform; Whelan et al., 'Real-time registration'.
[37] Rauber et al., 'Secure distribution'; Riley et al., 'Content-based retrieval'.
[38] Spector, Essays; Stevenson, 'Watermarks are twins'.
[39] Brockett, 'Aspects'; Hiary et al., 'Leeds Arabic'.

The key to our approach indicates the successful application of computer science.[40] Our aim was not to work directly with the image pixels (a bottom-up approach provided by widely used image manipulation packages). Instead, we tried to find a way to computationally model the image collection process, and thence seek to isolate the parts of interest. Figure 4 illustrates this: the image is a superposition of recto and verso information with the marks of manufacture, including other noise and damage. Under suitable assumptions we can systematically subtract away the recto and verso, leaving only the watermark and other mould marks, heavily but unavoidably corrupted by noise. This systematic technique sidesteps the often unrepeatable and ad-hoc properties of bottom-up work.

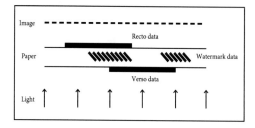

Figure 4: Cross-sectional image of paper with recto, verso and watermark data.

Figure 6 illustrates the entire mark after aggregation. In that figure can be seen evidence of chain lines and countermarking that are impossible to identify in all sheets of the original volume.

While extremely challenging, the data afford some small clues: on some sheets a small part of a watermark is evident. Figure 5a illustrates this. Importantly, this element is not visible on any but a very few sheets (it is in fact only necessary for it to be visible on one), and the pattern need not be complete. Such a fragment may be used with a statistical pattern-searching technique to locate, in the database of processed images, other instances of the same visible pattern that are very difficult or impossible for the naked eye to detect. Figure 5d shows an example of the integration of evidence culled from different pages: our algorithm has successfully identified and located the watermark.

At this juncture we are able to move into new territory: having some hundreds of instances of incomplete, noisy instances of the same pattern, we can superimpose them to recapture the entire watermark. The superimposition reduces the interference and noise properties. Thus we are able, automatically, to start with just a small fragment

---

[40] Boyle and Hiary, 'Watermark location'.

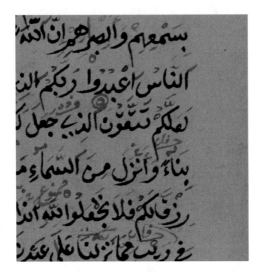

Figure 5: Illustrations of watermark location.[41]

Figure 5a: part of a straightforward scan of a page of the Mahdiyya Quran.[42] The recto script is carefully written; the image cannot show that the paper is thick and uneven. Many pages are damaged, discoloured or otherwise marked.

Figure 5b: the same material with a backlit scan. The presence of verso writing is now evident. Careful inspection shows some very faint watermark traces in the right hand margin.

---

[41] Boyle and Hiary, 'Watermark location'.
[42] Hiary et al., 'Leeds Arabic'.

Figure 5c: two watermark fragments that are easily seen on *some* pages, and so are easy to extract. These are a small part of the larger mark (to be seen in the left half).

Figure 5d: The identified locations of the fragments. Note that this evidence is hard to see in the backlit scan (significantly harder examples exist in the text), and that only a small part of the overall mark has been found.

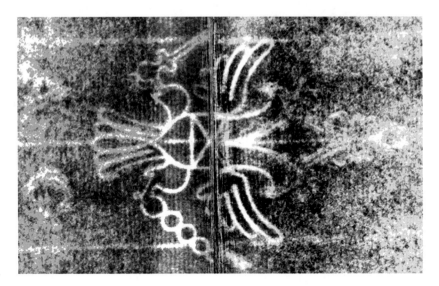

Figure 6: An aggregated watermark showing the double-headed eagle. Note the visible evidence of horizontal chain lines and at the left the countermark 'AG', a new discovery made possible by the computational approach that we have developed.

and recapture the entire pattern; the result is shown in Figure 6. As evidence of our achievement, this Figure shows a countermark, 'AG', which had not been identified in the earlier study by Brockett.[43]

A similar aggregation approach has demonstrated the ability to disambiguate twins, two near-identical marks in side-by-side moulds. Individually these are very difficult to distinguish by eye, but computationally the differences are clear. An example is shown in Figure 7.

This study has shown that properly applied computer science can allow us to see the invisible. The same techniques have also been used on other texts and shown to work reproducibly and faultlessly.[44] The contribution of computer science is a model-based approach that is robust because it is highly systematic, and founded in the physical process of image collection. The approach has borrowed algorithmic ideas taken from current artificial intelligence research, using techniques probably unavailable in consumer packages. This supports the analysis of materials that have to date been seen as deeply challenging,

---

[43] Brockett, 'Aspects', 1987.
[44] Hiary et al., 'Leeds Arabic'.

Figure 7: A superposition of trelune watermarks that exhibits twins. While very similar, the two superimposed watermarks are not identical. This is more evident in colour, and very evident in machine analysis.

moreover without requiring physical access to the original. Moreover, the cross-fertilisation has delivered useful results to codicologists and theoretical insights that benefit computer science.

Conclusions

We have considered the role of computer science and aspects of artificial intelligence outside their own disciplinary domains. Noting the very rapid pace of technical—notably hardware—development, we have also commented on the significant theoretical steps that the science takes. It has been illustrated that the application of such techniques, especially when married to faster machines and huge data storage that is Internet accessible, can bring large and sometimes unexpected benefit.

The domain of document analysis has been studied in the past but usually in the context of obviously remunerative applications such as postal address recognition or document surveillance for security purposes. Historical artefacts have received less attention partly because of lack of contact between the appropriate scholarly communities and partly because computer science is very frequently resource-driven and centres its attention on science and engineering activity. Nevertheless, increasing numbers of very fruitful applications of modern computer science techniques are in evidence.

Decreasing costs of hardware and software, together with ubiquitous Internet access, have brought a sudden wealth of data and new research opportunities to the desktops of all scholars. Much of this generates new insights and is therefore of great value, but often research does not draw on the contribution of professional computer scientists, thereby running the risk of under-exploiting research potential. We have illustrated several applications developed in computer science laboratories, in which the results obtained surpass those likely to be available by ad hoc, bottom-up approaches. The primary illustration has been the extraction and identification of watermarks from challenging source documents. The academic study of watermarks is far from new, but it has scarcely benefited from the attention of computer scientists. We have seen a model-driven approach succeed where the naked eye and ad-hoc techniques would have failed, with the successful identification of a hitherto unknown countermark and the separation of twins as our main evidence.

We consider this work an exemplar of cross-boundary collaboration that will be of benefit to all who participate. The relevant communities—in this case computer science and codicology—have thus far remained largely unknown to each other and have not yet developed a common language: Gants wrote in this respect of 'drawbacks [that indicate] the steep learning curve required for competence [and] the opaque language' used by computer scientists.[45] But he also noted that the adoption of computers by humanities scholars 'has altered the balance between collecting and analysing data'. The opportunities for mutual benefit are enormous, and the evidence presented here does little more than scratch the surface of future research potential. Further reinforcing interdisciplinary linkages and interaction between computer science and the humanities will perhaps create new hybrid scholars, and surely open up other significant avenues of work.

---

[45] Gants, 'Identifying'.

Acknowledgements

The work described in this paper was conducted within the School of Computing at the University of Leeds, and the Interdisciplinary Centre for Scientific Research in Music by the authors and Dr Kia Ng.[46] The support and contributions of ICSRIM, Hiary and Ng are gratefully acknowledged.

References

Ash, Nancy E. 'Recording watermarks by beta-radiography and other means', in *The book and paper group annual* 1 (American Institute for Conservation of Historic and Artistic Works, 1982).

Bentley, P., and D. Corne (eds), *Creative evolutionary systems* (San Francisco: Morgan Kaufmann, 2001).

Bernstein, The Memory of Papers project, http://www.bernstein.oeaw.ac.at, 2008. (Accessed: 12th December 2008.

Bower, P., *Turner's papers. A study of the manufacture, selection and use of his drawing papers 1787–1820* (London: Tate Gallery Publishing, 1990).

Boyle, R. and H. Hiary, 'Watermark location via back-lighting and recto removal', *International journal of document analysis and research*, 12, 1 (2009), 33–46.

Bradford, J., C. Needham, A. Bulpitt, and D. Westhead, 'Insights into protein-protein interfaces using a Bayesian network prediction method', *Journal of molecular biology* 362, 2, (2006), 365–386.

Bridgman, C.F., S. Keck, and H.F. Sherwood, 'The radiography of panel paintings by electron emission, *Studies in conservation* 3, 4, (1958), 175–182.

Bridgman, C.F., 'Radiography of paper', *Studies in conservation* 10, 1 (1965), 8–17.

Briquet, C.M., *Les filigranes. Dictionnaire historique des marques du papier dès leur apparition vers 1282 jusqu'en 1600* (Leipzig: Hiersemann, 1923).

——, 'Notice sur le recueil de filigranes ou marques des papiers, presented at the Paris Exposition in 1900', in *Briquet's opuscula. The complete works of C.M. Briquet, without 'Les filigranes'* (Hilversum: Paper Publications Society, 1955), 281–288.

Brockett, A., 'Aspects of the physical transmission of the Quran in 19th-century Sudan: Script, binding, decoration and paper', *Manuscripts of the Middle East* 2 (1987), 45–67.

CCH, Centre for Computing in the Humanities, King's College London, http://www.kcl.ac.uk/schools/humanities/depts/cch, 2008. (Accessed 27th November 2008).

Christie-Miller, I., Early paper project, http://www.earlypaper.com. (Accessed: 14th December 2008).

Churchill, W., *Watermarks in paper in Holland, England, France, etc. in the XVII and XVIII centuries and their interconnection* (Amsterdam: M. Hertzberger, 1935).

Codex Sinaiticus, The British Library et al., http://www.codexsinaiticus.org, 2008. (Accessed 27 November 2008).

Davidson, J., 'Measure for Measure: Exploring the Mysteries of Conducting', *The New Yorker* (21 August 2006), 60–69.

[46] ICSRIM, 2008.

de la Passardière, B., and C. Bustarret, 'Profil. An iconographic database for modern watermarked papers', *Computers and the Humanities* 36, 2, (2002), 143–169.

Dee, H., and S. Velastin, 'How close are we to solving the problem of automated visual surveillance? A review of real-world surveillance, scientific progress and evaluative mechanisms', *Machine Vision and Applications* 19, 5–6 (2008), 329–343.

DUIAH: Dutch University Institute for Art History, Florence—International database of watermarks and paper used for prints and drawings c. 1450–1800, 2008, http://www.iuoart.org/wmdb.htm. (Accessed: 12 December 2008).

Gants, David L., 'Pictures for the page. Techniques in watermark reproduction, enhancement and analysis', paper delivered at the annual meeting of the Bibliographical Society of the University of Virginia, McGregor Room, Alderman Library, April 1994, http://www2.iath.virginia.edu/gants/BibSocUVa/paper.html.

——, 'Identifying and tracking paper stocks in Early Modern London', *Papers of the Bibliographical Society of America* 94, 4 (2000), 531–540.

Gravell, T.L. and G. Miller, *A catalogue of foreign watermarks found on paper used in America, 1700–1835* (New York: Garland Publishing, 1983).

Gravell, T.L., 'The wizard of watermarks', *Du Pont Magazine* 84, 1 (1990), 4–6.

Haupt, W., 'Wasserzeichenwiedergabe in schwierigen Fällen', *Restauro* 87 (1981), 38–43.

Heawood, E., *Watermarks mainly of the 17th and 18th centuries* (Hilversum: Paper Publications Society, 1950).

Hiary, H., Digital extraction of archaic watermarks (PhD thesis, School of Computing, University of Leeds, 2008).

Hiary, H., R. Boyle, and K. Ng, The Leeds Arabic texts projects, University of Leeds, http://www.comp.leeds.ac.uk/arabictexts, 2008. With acknowledgements to the University of Leeds Library Special Collections and the School of Arabic and Middle Eastern Studies.

Huang, Y., M. Brown, and D. Xu, 'A framework for reducing ink-bleed in old documents', in *Proceedings of the IEEE International Conference on Computer Vision and Pattern Recognition* (Anchorage, 2008).

Hunter, D., *Papermaking. The history and technique of an ancient craft* (New York: Dover Publications, 1978).

ICSRIM, Interdisciplinary Centre for Scientific Research in Music, University of Leeds, http://www.leeds.ac.uk/icsrim, 2008. (Accessed 27th November 2008).

Inscriptifact, Image database of inscriptions and artefacts, University of Southern California, http://www.inscriptifact.com/, 2008. (Accessed 27 November 2008).

IPH International Association of Paper Historians, International standard for the registration of paper with or without watermarks, English Version 2.0, 1997, http://www.paperhistory.org/standard.htm. (Accessed 13th December 2008).

Koninklijke Bibliotheek (National library of the Netherlands), Watermarks in incunabula printed in the Low Countries, http://watermark.kb.nl/. (Accessed: 15 December 2008).

Loeber, E.G., *Paper mould and mouldmaker* (Amsterdam: Paper Publications Society, 1982).

Lowe, D., 'Object recognition from local scale invariant features', in *Proceedings of the International Conference on Computer Vision* (1999), 1150–1157.

Meinlschmidt, P., 'Original or fake?', *Fraunhofer—Research news* (July 2007).

Pavelka K., 'Review of *Puzzles in paper. Concepts in historical watermarks*', *Libraries and culture* 38, 4 (Fall 2003), 421–422.

Rauber, C., J. Ruanaidh, and T. Pun, 'Secure distribution of watermarked images for a digital library of ancient papers', in *ACM Digital Libraries '97. Proceedings of the 2nd ACM International Conference on Digital Libraries, Philadelphia, PA, July 23–26, 1997* (New York: Association for Computing Machinery, 1997), 123–130.

Repko, A., *Interdisciplinary research. Process and theory* (Thousand Oaks, CA: Sage Publications, 2008).

Riley, K., and J. Eakins, 'Content-based retrieval of historical watermark images. I: Tracings', *Proceedings of the International conference on image and video retrieval* (LNCS 2383; Berlin: Springer, 2002), 253–261.

Riley, K., J. Edwards, and P. Eakins, 'Content-based retrieval of historical watermark images. II: Electron radiographs', *Proceedings of the International conference on image and video retrieval* (LNCS 2728; Berlin: Springer, 2003), 131–140.

Rouse, R., untitled review of *The Problem of the Missale speciale*, *Speculum* 44, 4 (1969), 664–666.

Sahami, M., S. Dumais, D. Heckerman, and E. Horvitz, 'A Bayesian approach to filtering junk e-mail', AAAI Workshop on Learning for Text Classification, 1998.

Schnitger, D., and E. Mundry, 'Elektronenradiographie. Ein Hilfsmittel für die Analyse von Wasserzeichen und Miniaturmalereien', *Restaurator* 5, 1/2 (1981/82), 156–164.

Schnitger, D., E. Ziesche, and E. Mundry, 'Elektronenradiographie als Hilfsmittel für die Identifizierung schwer oder nicht erkennbarer Wasserzeichen', *Gutenberg Jahrbuch* 58 (1983), 49–67.

Schoonover, D., 'Techniques of reproducing watermarks. A practical introduction', in *Essays in paper analysis* (Washington: Folger Shakespeare Library, 1987), 154–167.

Small, C., 'Phosphorescence watermark imaging', in *Puzzles in paper. Concepts in historical watermarks. Essays from the International conference on the history, function and study of watermarks, Roanoke, Virginia* (Newcastle: Oak Knoll Books and London: the British Library, 2000), 169–181.

Sonka, M., V. Hlavac, and R.D. Boyle, *Image processing, analysis and machine vision* (3rd edn; Pacific Grove, CA: Brooks Cole, 2008).

Sosower, M., 'The Greek manuscripts written by Nicholas Turrianos in the library of Diego de Covarrubias (1577), Bishop of Segovia', *Codices Manuscripti* 41 (2002), 13–30.

——, *Signa officinarum chartariarum in codicibus Graecis saeculo sexto decimo fabricatis in bibliothecis Hispaniae* (Amsterdam: Adolf M. Hakkert, 2004).

Spector, S., *Essays in paper analysis* (Washington: Folger Shakespeare Library, 1987).

Stevenson, A., 'Chain-indentations in paper as evidence', *Studies in bibliography* 6, (1954), 181–195.

——, 'Paper as Bibliographical Evidence', *The library*, 5th series, 17, 3 (1962), 197–212.

——, *The problem of the Missale speciale* (London: The Bibliographical Society, 1967).

——, 'Shakespearian dated watermarks', *Studies in bibliography* 4 (1951), 159–164, http://etext.lib.virginia.edu/bsuva/sb/toc/sib04toc.htm. (Accessed: 13th December 2008).

——, 'Watermarks are twins', *Studies in bibliography* 4 (1951), 57–91, http://etext.lib.virginia.edu/bsuva/sb/toc/sib04toc.htm. (Accessed: 12th December 2008).

Stewart, D., R.A. Scharf, and S. Arney, 'Techniques for digital image capture of watermarks', *Imaging science and technology* 39, 3 (1995), 261–267.

Terras, M., 'Interpreting the image. Using advanced computational techniques to read the Vindolanda texts', *Aslib proceedings* 58, 1/2 (2006), 102–117.

Tomimasu, H., D. Kim, M. Suk, and P. Luner, 'Comparison of four paper imaging techniques. Beta-radiography, electrography, light transmission, and soft X-radiography, *Tappi journal* 74, 7 (1991), 165–176.

Toft, P., *The radon transform. Theory and implementation* (PhD thesis, Department of Mathematical Modelling, Technical University of Denmark, 1996).

UKCRC, Current grand challenges, http://www.ukcrc.org.uk/grand_challenges/current, 2008. (Accessed 27th November 2008).

van Aken, J., 'An improvement in Grenz radiography of paper to record watermarks, chain and laid lines', *Studies in conservation* 48, 2 (2003), 103–110.

van Hugten, H., 'Weichstrahlradiographie z. B. bei Papier', in *Vorträge des Symposiums Zerstörungsfreie Prüfung von Kunstwerken, Berlin, 19./20. November 1987* (Berlin, 1987), 43–49.

van Staalduinen, S., J. van der Lubbe, G. Dietz, and F. Laurentius, 'Comparing X-ray and backlight imaging for paper structure visualization', in *Proceedings of EVA, Electronic imaging and visual arts, Florence, Italy, April, 2006*, 108–113

Vander Meulen, D.L., 'The identification of paper without watermarks', *Studies in bibliography* 37 (1984), 58–81.

Vindolanda tablets online, University of Oxford, http://vindolanda.csad.ox.ac.uk/, 2008. (Accessed 27th November 2008).

Wenger, E., and V. Karnaukhov, 'Distributed database and processing system for watermarks', in Proceedings of EVA'04 Moscow, The State Tretyakov Gallery, Moscow, CD-ROM, 5 pages, 2004.

Whelan, P., P. Soille, and A. Drimbarean, 'Real-time registration of paper watermarks', *Real-time imaging* 7, 4 (2001), 367–380.

WIES—Watermarks in Incunabula printed in Espana, http://www.ksbm.oeaw.ac.at/wies/ (Accessed: 15th December 2008).

WZMA—Wasserzeichen des Mittelalters, http://www.oeaw.ac.at/ksbm/wz/wzma2.htm (Accessed: 15th December 2008).

Ziesche, E., and D. Schnitger, 'Elektronenradiographische Untersuchungen von Wasserzeichen in Inkunabeln', *Proceedings of the 19th International Congress of Paper Historians, Durham and Hertford, UK, 4–10 September 1988* (IPH yearbook 7, 1988), 209–223.

CONCRETE ABSTRACTIONS: ANCIENT TEXTS AS ARTIFACTS AND THE FUTURE OF THEIR DOCUMENTATION AND DISTRIBUTION IN THE DIGITAL AGE

Leta Hunt, Marilyn Lundberg and Bruce Zuckerman

Thanks to the recent availability of powerful technological tools that allow scholars to image and analyze ancient texts and artifacts in digital format and distribute their data worldwide, the concrete information about these texts and artifacts has never been better. However, it turns out that images of texts are more than just the depictions of concrete data that they would first appear to be. They also need to be seen and understood as abstract constructs that need to be conceptualized in terms of a theoretical model (or models) that conform to a methodology of access. In order to provide a solid foundation for intuitive access, one must establish and facilitate a strong relationship between the model and the 'real world'. The problem is that it is sometimes not altogether clear what the 'real world' is, and this 'reality-concept' shifts remarkably depending on one's perceptions. Drawing upon a number of illustrative examples from the records of the West Semitic Research Project and the InscriptiFact digital library from the University of Southern California, we will consider how one goes about developing these 'concrete abstractions' and the implications of this effort for the future of digital projects that deal with these witnesses to the past.

The reconstruction and decipherment of ancient texts has always been encumbered by serious and daunting obstacles. Particularly notable is the absence of reliable evidence that can serve as a confident basis for analysis, decipherment and interpretation of these inscriptions. In particular, a given reconstruction and restoration of a fragmentary text (and most of them are fragmentary) can rarely be verified with any confidence. Indeed, two scholars working independently are not likely to read and restore a text the very same way because the process is based on so little data that is reliably certain.[1]

---

[1] A telling example that exemplifies this point is found in the legal suit filed by Elisha Qimron against Hershel Shanks for violating the former's copyright of his

So what does this data usually consist of? The primary data for scholarly study of ancient texts traditionally has been provided by scholars (known in the trade as epigraphers) who have been able to gain physical access to the artifacts upon which the inscriptions were written. Based on their 'eye-witness' examination, they analyze these ancient texts, usually make detailed drawings depicting their significant data, transcribe them, translate them and supply supporting justification for the readings and their broader implications as well as other relevant commentary. This effort at epigraphic presentation and analysis is no easy task. The problem is that the artifacts rarely come down from antiquity in good condition; instead, they usually exist primarily as deteriorated fragments of parchment and papyrus, crumbling cuneiform tablets, abraded stone monuments and many other types of damaged media. Sometimes, for example, in the case of a deteriorated, blackened parchment scroll, it is even difficult to determine whether there is writing on the fragments at all, let alone what the texts may say. This has often led to considerable disagreement over what is actually written in given inscriptions; and, inevitably, much scholarly work is based upon what epigraphers reconstruct as the text. Nonetheless, scholars rarely have had the means to resolve disagreements, based on conflicting readings as evidenced in competing drawings and transcriptions, nor, as a rule, can they decisively resolve such questions by visual inspection of the inscriptions, themselves. Different pairs of eyes can and often do see different things. In fact, the way a given artifact is lit often prejudices how it will be interpreted—especially if it is has a depth dimension that is slight but significantly crucial for purposes of interpretation. (See Plate 11 for a typical example.)[2] Traditionally, a minimal set of illustrative pictures have often been taken, using, with varying degrees of effectiveness, the technologies of the time; but these often have only provided just enough information to fuel the debate and controversy over how the traces of a particular character

---

reconstruction of the Dead Sea Scroll known's as MMT (*Miqsat macase ha-torah*= 'Some precept of the Torah') by publishing it without permission. The court ruled in favor of Qimron because his reconstruction was deemed an *original* work, so hypothetical that it was nothing more (or less) than a scholarly-informed fiction. For a full discussion of the issues involved with this case, see T.H. Lim et al., *On scrolls* and the review of the same by B. Zuckerman at http://www.bookreviews.org/bookdetail.asp?TitleId=2951.

[2] For further discussion of this issue, see B. Zuckerman, 'Shading the difference', 233–252, 259–274 (illustrations).

or set of characters should be read. Frequently—and one might even say perversely—large issues of meaning with broad implications for the study of ancient culture, history and religion hinge on how a linguistically significant sign or set of signs is to be identified. Thus, it is not unusual to see more than one published article proposing mutually exclusive readings, all based on the same ambiguous image—with major issues trembling in the balance, depending on which reading is preferred.

As a result of these long-standing obstacles to progress in scholarly research, in the 1980s our group of scholars and technologists at the University of Southern California (USC) began to envision the concept of documenting ancient texts with a much more detailed comprehensive set of very high-quality, high-resolution images. The concept was to make it possible for all scholars, wherever located, to become 'eye-witnesses' via this substantially more expansive image documentation, so that arguments for readings and interpretation could be firmly based on more comprehensive data. If this goal could be achieved, it had the potential to revolutionize the way scholars and researchers of ancient inscriptions accomplish their work. The overall vision seemed as concrete and specific a goal as any that could be imagined: capture the tangible data (linguistically significant symbols incised on stone, inked on papyrus, parchment and pottery sherds, or impressed and stamped in clay, for example) in the optimal digital imaging format and make them broadly and easily available to the scholars who need them. However, the problems involved in achieving these complementary goals are just as significant as the problems involved in visualizing texts, noted above. For one thing, the technologies, both hardware and software, keep changing at a remarkably swift pace. Thus, as new and more powerful tools become available, they necessarily demand a constant and dynamic adaptation of methodology—a revisualization of what one needs to see in order to do the best scholarly work in reading, deciphering and interpreting ancient texts.

Moreover, it became clear that, in addition to the issues involved in developing technologies and methodologies for optimal image documentation, there are formidable obstacles to effective and intuitive online delivery of these images. As becomes all too obvious to anyone who has tried to find images of ancient inscriptions and texts by going to this-or-that internet accessible site, it turns out to be no easy task to find what one wants when one wants it in a quick, easy and intuitive fashion. To be sure, the image data may well be there in some manner, shape or form, but trying to get to it more often leads to frustration than to satisfaction. The trouble is that the *rationale* of access is often not well adapted to scholarly needs. In many cases, this happens because technologists who do not understand the nature and needs of scholarship in the humanities, in general, and the study of ancient texts, in particular, are the ones shaping the programs for access. In other cases, the culprits are the scholars themselves, who undercut their own efforts because they tend to think of an internet site as simply an electronic variation of a book or, more specifically, a text-edition/commentary. Hence, they unconsciously impose upon themselves and their sites what may be best described as 'bookish' limitations that tend to impede easy access to their data.

What we have begun to recognize is that images of ancient texts are more than just the depictions of concrete data that they would first appear to be. They also need to be seen and understood as abstract constructs that need to be conceptualized in terms of a theoretical model (or models) that conform to a methodology of access. In order to provide a solid foundation for intuitive access, one must establish and facilitate a strong relationship between the model and the 'real world'. The problem is that it is sometimes not altogether clear what the 'real world' is, and this 'reality-concept' shifts remarkably depending on one's perceptions. How one goes about developing these 'concrete abstractions' is the chief concern of this essay.

Fundamental to the issues of documentation and distribution of ancient text is the need to facilitate reliable reproduction and reconstruction of ancient texts in a distant (i.e, internet) environment (that is, with the focus not on the actual artifacts but rather on high quality digital reproduction of the artifacts and their distribution). Thus, ultimately, there should be a means to accurately reproduce in an internet environment what was written in a variety of contexts (differing approaches to digital scanning, varying film types, strategies for lighting, etc.); to compare images and effects side-by-side, so that scholars can view the image data from different perspectives at the same time in real time; to reproduce the effect of changing the lighting angles similar to an eye-witness being able to tilt a given artifact or move it closer to a light source (for example, by presenting different images captured from different light angles, or preferably to facilitate changing the light angle in real time); in the case of three-dimensional artifacts, it would be helpful if the various surfaces could be viewed in context so that writing that wraps around the edges or from obverse to reverse, for example, could be viewed (or preferably to rotate the object's reproduction with the capacity to zoom in on any surface); to toggle on and off mathematical enhancements (preferably in real time) of the visualization so that subtle changes in the marking can be perceived more easily with the human eye.

The West Semitic Research Project (WSRP) and the InscriptiFact Project at USC were initiated as a collaboration among scholars and technologists to further this vision. The aim of WSRP is to research and employ, typically through field documentary projects, the most effective imaging technologies to record in high-resolution images of a broad variety of ancient texts (especially from the Near East) crucial to scholarly understanding of ancient history, religion and culture. The goal of InscriptiFact is to provide fast, reliable and intuitive access to images of these ancient texts acquired through WSRP and its collaborators to all interested parties, world-wide.

This study will discuss the accomplishment of the vision both in terms of imaging technologies and implementation of optimal distribution of the images. It will further consider some of the most important and most recent imaging technologies that significantly improve the reading, analysis, reconstruction and decipherment of ancient texts and how their images can be effectively and intuitively distributed over the internet, using InscriptiFact as a model. More specifically we will consider the implementation in an online context of some of the most important viewing features and real-time functionalities that facilitate better overall understanding of ancient texts. In sum: we will discuss the status of our vision and the future we hope it is shaping.

Over the nearly three decades during which WSRP has initiated field documentary projects, computer imaging technology has provided increasing opportunities to visually record ancient texts more effectively. New techniques are constantly evolving, and improvements in imaging hardware are continually tested and are being applied as appropriate. Each of these has brought with it new challenges in imaging methodology. However, from the beginning, three foundational and complementary goals have been recognized as essential to a strategy to produce the best possible image documentation: quantity, quality and variety. If each inscription could be documented with an expanded series of detailed, high-quality images, employing a range of technologies (for example, digital scans, a variety of film types, light sources, levels of magnification and so forth), always with a concern to preserve in detail every potential aspect of the inscription, such images, when made freely and intuitively available to scholars world-wide, could better serve as primary data for purposes of resolving questions of interpretation than had ever been previously possible. Our group believes that the key to success in this endeavor has been to self-consciously determine how to reinforce detailed scholarly knowledge of the inscriptions and their cultural environment with experience and expertise in the use of the most advanced imaging techniques. The resulting images can provide a new and far more reliable foundation upon which to base serious scholarly research.

Imaging technologies and functionalities that significantly improve visualization, decipherment, analysis and reconstruction of ancient texts

In previous years WSRP typically documented ancient texts with large format (4x5 inch) films that were subsequently digitized, employing high-resolution scanners. While such scanning continues for purposes of digitizing older photographic plates (some as much as a century old), a number of years ago we began to rely almost entirely on 'born-digital' technologies for ongoing work. Current large-format viewing cameras combined with high resolution scanning backs can deliver more detailed and color-accurate images in visible light color and superior resolution in infrared than have ever been possible before. For example, visible light capture (see Plate 12b) of a Dead Sea Scroll in high resolution, while not showing the inked letters very well, enables close examination characteristics of the parchment itself, including important information relevant to the artifact, such as hair follicle patterns and surface degradation. Infrared capture (at the appropriate narrow bandwidth; see Plate 12b on the right below) reveals the writing itself so that the text can be analyzed, deciphered and interpreted. Neither image should be done to the exclusion of the other, for infrared capture cannot provide the same kind of information about the skin itself upon which the inscription is written (important, for example, for purposes of conservation) while the readings are best verified based on infrared imagery.

Note in this regard that the technologies available for infrared capture have dramatically improved since the advent of digital imaging. Contrast the infrared image captured in 1988 with large format high-speed infrared film (at that time the best available means of doc-umentation), with the infrared image captured in 2008 with a high-resolution digital scanning back in Plate 13. Notice that the older image is grainy and shows significantly less background detail due to the intrinsically limited resolution of the film,[3] while the newer image, capitalizing on the far greater resolution of a scanning back, achieves a far greater degree of image sharpness than was previously possible.

New methods of capture have evolved, focused on examining the surface texture of an inscription. This is especially useful in the analy-sis of incised markings on hard surfaces but can often even pick up far more subtle physical dimensions, including the thicknesses of ink strokes on a manuscript or paint strokes in an illuminated design. More recently, such techniques have shown great promise for the study of texts on parchment and papyrus by facilitating analysis of scribal ductus, brush technique and in reconstruction of fragmentary texts based on the characteristics of hair follicles on parchment, the grain of papyrus or common patterns of damage. One especially useful technology is 'Reflection Transformation Imaging' (=RTI). RTI image-objects are typically captured by placing the artifact within a dome upon which are mounted lights in an evenly distributed array of posi-tions and heights and with a camera in a fixed location directly above the subject artifact (see Plate 14). Numerous images are captured automatically (usually, 32 in the domes currently used by WSRP), each with the light source at a different height, position and angle. These images are then synthesized to make a single, master image-object. A specialized but easily downloadable viewer allows the user to change the apparent direction of the light source by moving a mouse-driven cursor around the image in real time. The technology also allows the image to be mathematically enhanced in a number of useful ways. For example, in the 'specular enhancement' mode, the image of a given surface can be made to appear highly reflective, as though it were cov-ered with molten silver, bringing great clarity to the slightest details of an inscription (see Plate 15).

Field technologies are being developed that facilitate the image-cap-ture for texts and artifacts in challenging field conditions. For exam-ple, techniques have been developed to capture RTI images-objects of

---

[3] See B. Zuckerman, 'Bringing the Dead Sea Scrolls back to life', 188–189.

inscriptions on large scale monuments (for example, cliff walls) that do not require the use of a dome (which is better suited for smaller objects in a laboratory environment). Such field technologies are very much in the beginning stages of development but offer great promise for facilitating visualization of ancient inscriptions quickly and effectively in a manner not previously possible.

New methodologies and equipment are capable of restricting potential harm to light sensitive artifacts by limiting their exposure to light. For example, the 'slider' technique allows an image to be progressively advanced across a small light source while synchronized to the line scan of a scanning back, thus limiting the total amount of light applied to the artifact at any given time. Such an approach also is ideal for controlling lighting effects over an object of significant length and/or width (see Plate 16).

An especially exciting use of digital technology involves the study of cylinder seals. Cylinder seals are small cylindrical shaped stones that have intricate carvings on their curving faces. When these stones are rolled out on wet clay, they produce a continuous impressed image. Such seals were used in antiquity as the ancient equivalent of a signature to endorse various types of transactions written in cuneiform on clay tablets. A collaborative project between the University of Illinois (Urbana-Champaign) and USC, in which undergraduate researchers did the principal work, employed an innovative WSRP adaptation of digital panoramic photography to capture the actual carved surface of 62 cylinder seals in 360 degrees as a continuous, flat image. This new technique facilitates the analysis of cylinder seals (or other similar cylindrical shaped artifacts) with a level of sophistication and detail that was previously impossible when the only means of reference was the 'roll-out' of the image on clay. Tool marks and even minute corrections made by the craftsmen are revealed in detailed images (see Plate 17).

Within this context of continuous improvement, WSRP has developed standards, principles and best practices for digital documentation of various kinds of ancient texts, while approaching each field project as a new opportunity to document a text in the most effective way, creatively using whatever technologies are appropriate and applicable within the limitation of a given environment.

But documentation of images is only half the battle. Distribution of the data, once collected, is also of essential importance. Let us now look at the planning for an optimal means of distribution of these images to scholars, and the implementation in an online context of some of the most important data relationships and real-time functionalities that facilitate reconstruction of ancient texts.

Optimal Distribution of Images

As WSRP scholars continued to develop methodologies for establishing the optimal means of documenting ancient texts, especially during the course of a number of field projects, the need for an effective means of distributing their data to scholars and researchers world-wide came to the forefront, especially as the facility of the internet made this progressively more feasible. Initially, a WSRP website was developed where a limited number of images were posted (see www.usc.edu/dept/LAS/wsrp) and a searchable catalogue of a wider number, not directly available through the internet, was presented. But, as the site grew and the demands of scholars became more significant, it was recognized that a more elaborate and all encompassing image-database was necessary to fulfill research and educational needs. The initial site was simply not sufficient for the vision of making the images intuitively and easily available world-wide. The team regrouped to ponder the distribution component of the vision more seriously.

A discovery process was introduced beginning in 1998 to develop and document the scholarly vision of optimal display and distribution of images. The team of scholars and technologists began to consider a rationale of access that would be appropriate especially for the primary audience of scholars who study ancient texts. As was mentioned previously, in order to provide a solid foundation for intuitive access, one must establish and facilitate a strong relationship between the model and the 'real world'. The problem is that it is sometimes not altogether clear what the 'real world' is, and this 'reality-concept' shifts remarkably depending on one's perceptions. That is, a fundamental difficulty in implementing a digital library to serve scholars and researchers of ancient texts in an effective and optimal fashion is that concepts foundational to related research function on an abstract level. They do not function in the world outside of their scholarly venues, and they are often not readily understandable nor are they easily communicated to the non-specialist. From a technological perspective, this presents a significant potential for misunderstanding and error during the planning and analysis phases of developing a digital library, resulting in equally significant implementation errors that present obstacles to intuitive access. We recognized early on that it was essential that our distribution application, the InscriptiFact Digital Library (ISFDL), capture these abstractions and transform them into useful, concrete functionalities.

In the case of the InscriptiFact, the nature of the relationship between ancient texts and the artifacts on which they are inscribed proved to be a fundamental issue to defining an intuitive rationale for intuitive access. The problem faced by our team of scholars and technologists is conveniently illustrated by the story of how 'InscriptiFact' got its name. The scholars in the initial requirements sessions were attempting to clarify to the engineers how they define an 'ancient inscription', that is, how one inscription is differentiated from another, and how the concept of a given inscription is related to the physical artifact(s) (clay, stone, animal skin, papyrus, etc.) on which it is inscribed.

Explaining this relationship precisely is not as direct a task as it would appear, since a single ancient inscription may exist across several physical artifacts and, conversely, there may be more than one inscription on a single artifact. Another problem we faced in the case of InscriptiFact, is that there was a conventional expectation for a digital library to be organized around the abstract concept of a cohesive, 'intellectual work'. However, defining such an 'intellectual work' in terms of ancient inscriptions as well as the artifacts, upon which they are written, is not quite the same as defining a modern 'intellectual work'. For one thing, an ancient text often has unknown, uncertain or questionable origins, and there may be a good deal of ambiguity regarding the extent and organization of the text. For example, as an ancient corpus of inscriptions is studied, fragments once thought to be a part of one text may be later considered to be part of a different text, or a given text may be subdivided into several texts as a result of the ongoing research process. Indeed, one scholar's reconstruction of a text, within the context of a wider corpus of texts, may be quite different than that of another scholar.

Predictable dubiety in the accurate reconstruction and reading of an ancient text should be reflected in the data model of the digital library intended to serve scholars who study ancient texts. That is, if a digital library of ancient texts must be organized around the concept of a cohesive written intellectual work, one must have the fundamental capability of determining what *is*, in fact, that intellectual work. While such foreknowledge is a given for most intellectual written works of the modern era and thus also tends to be a given in most indexing systems employed by libraries, this assumption is frequently inappropriate if not irrelevant when applied to ancient texts.

Moreover, it is rarely possible to determine whether a given text is an original work or a copy, in whole or in part, of some other, earlier text. Consider, as examples, the two Ketef Hinnom Amulets, dating from the seventh to the sixth century BCE, originally found among the debris in a burial chamber just outside the city walls of Jerusalem.[4] These inscriptions invoke, each in a slightly different version, the prayer, well-known in Judaeo–Christian tradition as 'the Priestly Benediction' and cited in Numbers 6:24–26 ('May the LORD bless you and keep you...,' etc.). These objects (specifically, two tiny ribbons of silver lightly incised in a minute and difficult-to-read paleo-Hebrew script and then originally rolled up like tiny scrolls) are the earliest known artifacts to cite text also found in the Hebrew Bible. But there is hardly any certainty regarding how to describe these texts more precisely than this. Are the texts on these two amulets best viewed as the earliest *direct* citations of a biblical verse, or are these texts—especially since they vary slightly from each other and also from the form of the prayer found in Numbers—better seen as reflections of a prayer-tradition in the sixth century BCE that was extra-biblical or possibly even pre-biblical? The prayer inscribed on each amulet appears to be quoted in a broader and somewhat uncertain context that may include phrasing from other biblical texts (or at least texts that find their way into the Hebrew Bible). How does one characterize the relationship between these possible other quotations, the Priestly Benediction and the unattributable and uncertain texts of which they are a component? As a result of these ambiguities, description of the intellectual work of these ancient texts cannot be easily separated from the physical objects upon which they were inscribed.[5]

Issues of this nature led one engineer to query with some exasperation: 'How do you denote the inscription itself that exists on or

---

[4] For a full and recent discussion of these texts, see Gabriel Barkay et al., 'The amulets from Ketef Hinnom', 41–71 (with a companion CD-ROM containing extensive illustrations).

[5] For more information on the problems of modeling the data in InscriptiFact, see Leta Hunt et al., 'InscriptiFact', 3.

across an artifact—or more than one artifact—but is not the artifact itself, may be only indirectly related to the artifact(s) on which it is inscribed, but cannot be considered separately from the artifact?' How indeed! A scholar responded a little defensively and a bit more than half-humoredly: 'Maybe we should just call it an "inscriptifact".' Almost everyone in the room immediately recognized that this just-coined term had potential to encompass and describe something that we could not quite so easily define before: 'facts' embodied in an 'inscription found on an artifact' (thus the capitalization of the 'F': 'InscriptiFact'). Granted, an 'InscriptiFact' does not exist in the real world, but this abstract characterization of concrete remains turned out to be basic to an understanding of the way philologists, epigraphers and others who study ancient inscriptions approach the study of these texts. Hence, it provided one of the foundations for the data model ultimately implemented in the image library that was endowed with its name—InscriptiFact.

Thus, a set of related questions peculiar to the design of an appropriate model of texts and artifacts had to be answered: for example, what is the relationship, and what is the value of the relationship, between ancient texts and inscriptions and the artifacts upon which they are written? Further, what relevance does this relationship have to providing access to images of ancient texts? The answer to this question is both simple and complex: it depends entirely on one's perspective.

Let us consider potential options: if an InscriptiFact user is a theologian, teacher or student without in-depth professional training in the field of ancient inscriptions and artifacts, her or his aim will likely be to discern the broad meaning of a text (for example, an early biblical manuscript) in terms of its relevance to the study of history, religion and culture in a modern context. The relationship of the text to the artifact is viewed simply as that of an intellectual work to the medium on which it is written (similar to the relationship of the Constitution of the United States to the parchment, upon which it was inked). The relationship is of limited significance, except perhaps to illustrate the text's authenticity as something that actually existed in antiquity.

If an InscriptiFact user is an art historian, her or his focus will be on the non-textual but human-created elements found on a given artifact. Once again, the inscription itself may be of some significant importance to the art historian, but primarily in terms of what it can do to contextualize the more purely artistic and symbolic elements to be found on a physical object. The relationship of the artifact to the art it displays is obviously something that the art historian considers closely, since such a rendition is necessarily dependent on and has a close relationship to the medium upon or through which it is produced (for example, a statue or bas relief carved in stone, a crude drawing scratched next to graffiti on a cliff side, the imprint of a god figure in a clay sealing or bulla, an inked drawing of an animal on a papyrus).

If an InscriptiFact user is an archaeologist, his or her focus will primarily be on the artifactual aspect of a given inscription. That is, an archaeologist views an inscription first and foremost in terms of the physical object upon which it is inscribed, the physical location of that object (for example, in what stratum of settlement it is to be located on an ancient tell) and its relationship to other objects (for example, those that were found nearby, those both on and off site that are typologically similar). Hence, for the archaeologist the writing, while obviously of some intrinsic importance, is not the main thing. It is simply a part (perhaps even a minor part) of a bigger picture of the assemblage of artifacts to be assessed and classified in context found in the various loci of an archaeological dig and in the broader context of the material culture of a given civilization.

The team desired that the ISFDL be useful to a variety of potential users and wanted it to be sufficiently flexible to encompass all their needs.[6] Nonetheless, it is for the needs of the epigrapher, especially

---

[6] For the theologian, InscriptiFact serves as a facilitator of illustrations. A rabbi may wish to see a good picture of the earliest inscription (an Aramaic papyrus from the fifth century BCE Egypt) to mention a Passover ceremony. An educator may want to display a good picture of a Dead Sea Scroll (for example, the Temple Scroll of the first century BCE from Qumran) for her or his college lecture on the religion of early Judaism. A high school student may wish to have a nice picture of a cuneiform tablet (for example, economic record of the second millenium BCE from Ur III) to include in her or his high-school report on the beginnings of literacy. For the art historian, InscriptiFact can be of considerable value by providing high-resolution images for detailed analysis of an artistic work or even how the work was manufactured (as in the case of a cylinder seal in the illustration above). Moreover, InscriptiFact deliberately includes a number of non-textual artifacts that feature outstanding artistic work of particular interest to the art historian. For the archaeologist, because InscriptiFact delivers images in very high resolution, important information about the physical aspects of a given inscribed artifact, the physical artifacts are extraordinarily well illustrated (for example, the grain of the individual strips that compose a papyrus page; the texture, color and grit content of the clay sherd of an ostracon; the hair follicle patterning of an animal skin scroll) that may not otherwise be available. Moreover, InscriptiFact supplies detailed cataloguing information that will allow an archaeologist to reference

in her or his role as a philologist (i.e., someone trying to discern the meaning of a text) for which InscriptiFact was to be optimally designed. While the artifact is of some intrinsic importance to the epigrapher, it is not the main thing; rather, the artifactual aspect of the inscription is considered by the epigrapher primarily in terms of the clues it possesses that will help guide her or his understanding of the date, cultural environment and—above all else—the reading and meaning of the text in and of itself. But, the relationship goes much further than that. One of the differences between ancient texts and modern texts is that, comparatively, there are not very many of the latter for an epigrapher to study. Where an archaeologist will have numerous (even millions) of artifacts to study from a single site and an art historian likely will have a fair range of materials (for example, the aesthetic aspects of architectural structures) to work with, an epigrapher/philologist will count her or himself lucky to have just a few—especially in the Syro–Palestinian milieu which is the central academic focus of the ISFDL. Inevitably, this paucity of data leads to the need for extensive reconstructive (and imaginative) analysis along the lines noted above and depends substantially on clues that can be ascertained from the artifact itself.

The needs of epigraphers encompass particular concerns. For example, text genres in antiquity are of special significance. Each of these may encompass a broad variety of artifacts. Some of these texts are similar to templates that are used in particular circumstances, such as contracts or legal documents in a given culture. Other forms are used, for example, for curses, blessings, and warding off evil. There are texts that although closely related, have some notable differences that add significant value in the analysis of the text—especially the history of the traditions that they may exemplify. There are texts, such as palimpsests in which two texts share the same artifact, one on top of the other; that is, the artifact (usually parchment or papyrus) on which one text has been written has been 'erased' ('palimpsest' literally means 'scraped over' in Greek) and reused for a different text. One example of this is Codex *Syrus Sinaiticus* from St. Catherine's Monastery on the Sinai Peninsula, in which the earliest copy of the gospels in Syriac has

---

its physical context at a given excavation, as well as its typological placement relative to other similar artifacts. Such information can also be of considerable value to conservators and preservationists as well.

been written over with a text that scholars have entitled 'The Lives of the Female Saints'. In such cases, the epigrapher needs to judge how the two texts are related and why one was erased in favor of the other, not to mention resolving the often formidable problems of studying the text that has been scratched out.

What is actually necessary for the epigrapher to study is not just the text as an isolated bit of data, or the artifact as single physical object, but rather more the 'instance' of that particular text: the text, the artifact, the find site (if known), the means of communication (script; language; type of ink used, writing instrument, etc.), the relationship of the particular text to other texts, the cultural similarities that can be discerned between the scribe and the culture in which the text was located (for example, was the scribe on a journey from one place to the other, as was the case with the graffiti-like inscriptions on cliff walls in the Wadi el Hol, Egypt—currently the earliest attested alphabetic inscriptions),[7] and so forth. Every element of the instance that can be discerned, classified and related in terms of the greater historical and cultural picture is of critical importance because there are so few examples of ancient texts and thus so few clues remaining to enlighten us. The perspective of the epigrapher is that the meaning of the text is inextricably related to the instance in which it was written. It is this perspective, in particular, that InscriptiFact is meant to address.

Thus, the ISFDL model is based on a text as defined by its instance, including the physical artifact(s) on which it is written. A given text within the context of a separate and distinct instance is regarded as a separate text.

A responsive application will have an underlying data model that reflects the 'real' (albeit abstract) world of the primary clientele. This can present a challenge when two groups that work with the same data have differing perspectives. On the other hand, a one-size-fits-all approach may well end up being a flawed, 'one-size-fits-none' strategy.[8] For example, the Online Cultural Heritage Research Environment (OCHRE) database of the Oriental Institute, University of Chicago, is designed around the perspective of archaeologists and historians, while the ISFDL is, as noted, designed around the perspec-

---

[7] See John Darnell et al., *Two early alphabetic inscriptions.*
[8] See note 4 for an article reference to discussion of application development strategies as they relate to specific versus one-size-fits-all approaches to application development.

tive of epigraphers and philologists. The two databases contain largely differing content about the same corpus, the Persepolis Fortification Archive, with OCHRE containing archaeological information as well as editions of texts in progress, and InscriptiFact containing high-resolution images with extensive cataloguing data but no transcriptions, translations or commentary. The aim of ISFDL is to supplement the aims and intents of OCHRE by supplying the best visual data (and the context of other similar images of inscriptions) to facilitate reading(s) and interpretation(s), but it does not try to do what OCHRE intends to do—specifically to set texts in a more detailed archaeological and historical context.

It would obviously be useful for users to be able to move from one database to the other with ease. For this reason, requirements are in the process of being developed that will enable ISFDL users to move easily and intuitively from InscriptiFact to OCHRE (as well as other internet-accessible sites) and *vice versa*. Such interaction requires a 'translation' of sorts between data models and will be increasingly necessary for the ISFDL as it continues to develop and expand—not to mention other digital-image libraries that require similar interactions in order to be of most effective use to a wide range of users. In this fashion the particular perspectives and concomitant concerns of theologians, educators, students, archaeologists, art historians, epigraphers and philologists, as sketched out above, can be better served by a network of specialized but complementary repositories of data that allow for the fullest range of use.

As noted, the model and the basic features of InscriptiFact always takes into account the concrete abstraction embodied in the complicated relationship between text and artifact. While the underlying organization of the database revolves around the concept of 'text', consideration of the particular instance of a text on an artifact is directly relevant to how philologists and epigraphers think about that text/ inscription. Thus, search parameters refer to both artifact (the concrete) and text (the abstract).

Let us consider this point in specific reference to the search categories the ISFDL employs. A search for a given text is facilitated by any one of a combination of parameters including: 'Corpus of Texts', that is, a convenient academically determined grouping of artifacts and texts that are understood to have something in common within the scholarly field (for example, the 'Dead Sea Scrolls'). 'Medium', on the other hand, refers specifically to the 'artifact', that is, the material composition that the text/inscription is written on. 'Find Site' again pertains to the artifact, but also a text written at a given location, within a particular cultural milieu. 'Time Period' (defined by any one of three chronologies depending on the area of concentration, currently including Mesopotamian, Egyptian and Syro–Palestinian—but with more planned as the purview of InscriptiFact broadens) provides information about the artifact, which contains an inscription, but may or may not let one know when the text was originally composed. For example, the Dead Sea Scrolls are a group of scrolls and fragments of scrolls that were found in a series of caves at Qumran, along the shores of the Dead Sea. A parchment fragment of Genesis found in Cave 4 of Qumran can be dated to the Roman period, but the text, the intellectual work, of Genesis was written long before. What is presented in InscriptiFact is one instance of that intellectual work, dating to a particular time. 'Language' points to the text, or conceptual entity, rather than the artifact itself, but 'Script' relates to both artifact and text, in that it refers to the way a particular text is represented on a material object, and the writing system itself affects the way the conceptual nature of a text is expressed. 'Repository' designates an institution where given artifacts are located, and a text may be divided between one or more Repositories. Another option in the ISFDL is a search by keyword, either by itself or in combination with any of the above.[9]

In the process of reconstructing a text, varying identification formats (sigla) may be employed at different times for various scholarly editions. As a result, it is not unusual for there to be a number of competing identification systems. In response to this reality, InscriptiFact has been structured to recognize *any* of these identification systems, so that a scholar can request a text via whatever system that she or he is most familiar with and retrieve the relevant texts, by specifying a 'Text and Publication Number'.

The manner in which an InscriptiFact search proceeds also takes into account the 'artifactual' nature of a text. Once search parameters are chosen, the user is presented with a list of texts that fit these parameters. Depending on the characteristics of a selected text, the user may be presented with further means to refine selection of images. For example, if the text is lengthy, the user may be presented with

---

[9] See note 3 above for an article reference to more information on InscriptiFact search.

a list of text divisions (such as columns or manuscript pages). Such columns do not necessarily represent meaningful breaks in the text, but rather show how the text was physically laid out on the artifact. In cases where a researcher wants to view just one physical area of a text, InscriptiFact facilitates a 'text spatial search'. A reference view of the text (for example, a front, back or, where applicable, an edge) is presented, and the user can use his or her mouse to define a box of any size and dimension over any part of that reference view. Once the user clicks on the 'Go' button (see Plate 18), the images that are retrieved are only those that intersect with the box the user has defined. With the quantity and variety that are often involved in WSRP's typical documentation, a global search of a given text might require a user to consider many images—sometimes hundreds—whereas, the text spatial search component allows her or him to go quickly to a few relevant images (perhaps, for example, a single letter or group of letters of crucial importance to the interpretation of the text).[10]

In structuring InscriptiFact it became necessary to take into account the fact that a text (for example, a coherent narrative) might no longer exist on a single, intact, artifact. Thus, InscriptiFact, using the text as an organizing principle, brings together all the parts of an artifact on which a text is written, and allows them to be viewed together regardless of the collection or repository in which the various fragments may now exist (see Plate 19). However, the artifactual aspect of a given instance of a text is still taken into account, in that different instances of a text, written on different artifacts, are each catalogued and indexed as separate texts, the principle being to always present a text within the context of its particular instance. To go back to our example of the Dead Sea Scrolls, fragments of several different copies and versions of the biblical book of Genesis are represented in the finds from the various caves at Qumran. Each copy and version must be considered as its own text, because, at that period of time, each manuscript might have slight or potentially larger variations in the spelling or the wording of the same chapter of Genesis, and these variations are of great interest to epigraphers and philologists. Moreover, varying historical and cultural conditions may also impact a given text so that it is best viewed as an entity, in and of itself, even though it may be obviously related

---

[10] For more detail on the InscriptiFact spatial search, see Leta Hunt et al., 'Nongeographic spatial search'.

to other versions of the broader text tradition. This can be compared to the library practice of cataloguing each modern translation of the Bible as a separate book.

The most recent InscriptiFact viewer facilitates Reflection Transformation Images (RTI), which as noted above, enables scholars to change the direction of the apparent light source by moving a mouse-controlled cursor. The InscriptiFact viewer also facilitates the toggling on and off of RTI-enhancements of the sort discussed above (for example, specular enhancement). For example, in Plate 20, the same digital object is loaded into two side-by-side frames, with left frame showing the artifact in its natural color and appearance and the right frame showing the same artifact in specular enhancement mode. Note that in either case the apparent light source can be changed by moving the mouse controlled cursor around either image in real time in InscriptiFact.

InscriptiFact also facilitates the side-by-side comparison of RTI images with conventional images (Plate 21). Also, InscriptiFact facilitates viewing the reproduction with more than one light source, allowing scholars to adjust the dual light sources to the best advantage of the particular text section they are reading (Plate 22).

The ISFDL strategy of organization is designed to facilitate scholars moving quickly and intuitively through the range of images needed for optimal epigraphic and philological study. As noted above, there is a tendency in many image archives to present things as though they were still in the standard presentation form of the past—the book, or, more specifically, the codex. In such presentations, the display seems two-dimensional or 'bookish' with many options often displayed (one might even say crammed) on a single page. The rationale for this approach was understandable when space was defined by the demands of the printing process on paper, and there was a natural economy in putting more data on a page rather than less. That is, it is less expensive to print and bind a book with fewer pages. For the same reason, illustrations (always an added expense in a paper environment) were usually kept at an absolute minimum in a book in order to keep costs down.

However, in an electronic environment, display space is a cheap commodity. For this reason, the ISFDL always aims to give its users as broad a range and variety of high quality images as are available. But this, in turn, poses a potential problem for the users since such a welter of images and the wide number of choices and options available can be confusing. Our strategy in regard to this concern is simple and direct: less is more. That is, in order to address this concern, ISFDL deliberately constrains the choices a user can make on a given display page to relevant options based on the user's previous selections. Instead of confronting numerous choices on a single screen, the user is directed to drill down from one screen to another as the specific search dictates—with the avenues of choice defined by scholarly editors who catalogue texts and, in so doing, predetermine the available options based on the characteristics of the particular text selected. (See above discussion regarding 'List of Text Divisions' and 'Text Spatial Search'.) The result is that users can quickly and intuitively zero-in on the small number of images relevant to their research as opposed to the potentially hundreds of image options that are fully accessible. The idea is to give the user many options, but not too many options *at once*.

Finally, the way an inscription is written onto an artifact dictates the imaging methodologies that are required to make the text clear as it is represented on the artifact. It has been our effort to image inscriptions in such a way as to show, where appropriate and useful, the nature of artifact surfaces on which inscriptions are written. This is accomplished by using a variety of light sources to show surface texture (see above) or lighting in such a way that one can see into grooves or incisions that make up meaningful symbols. InscriptiFact allows images from these different imaging methodologies to be compared side-by-side.

An InscriptiFact user was questioned recently at a conference as to whether the ISFDL has actually made a difference to scholars and researchers. More specifically, the question posited was whether a digital library such as InscriptiFact is simply an interesting technological instrument that can be used to gain temporary momentum for specified projects, or whether it plays a fundamental role in pioneering and ongoing change in its field of study. The user's response was that 'it has revolutionized our field' (the field involving the reconstruction, reading and interpretation of ancient texts).[11] This indeed is our intent: for

---

[11] An unpublished usability study, funded by the Andrew W. Mellon Foundation and implemented by an independent third party, concludes that InscriptiFact is regarded by more than 90 per cent of users and potential users as having a highly significant impact on the scholarly practice, ability to teach, and ability of students to learn in its targeted fields. Bruce Zuckerman et al., 'InscriptiFact usability study'. Some of the usability study results are also published in, Leta Hunt et al., 'Inscripti-

InscriptiFact—and the scholarly and informational logic upon which it is based—to revolutionize the way scholars and researchers of ancient inscriptions accomplish their work.

Conclusion: Status of the Revolution. Envisioning the Future

As the discussion above has intended to demonstrate, a technologically driven revolution is transpiring before our eyes. An increasing number of scholars are learning advanced image documentation techniques and considering them fundamental to research and study of a wide range of ancient texts. Students, even at the undergraduate level, are learning image documentation techniques and facilitating scholarly study of texts and artifacts in ways that have never been previously available.

Key to progress is increasing reliance on personal computers. Scholars and students need all the memory and processing speed they can get to fully utilize the image objects that are currently available. Current limitations in file size, (for example, in RTI capture) limit the surface detail that can be provided in RTI images. But issues of this nature will soon be overcome. Indeed, the possibilities for new documentation technologies are far-reaching and open-ended and depend only on the creativity and collaboration of engineering researchers and scholars. Field technologies need to be further developed to facilitate advanced capture in challenging, remote terrains. Three-dimensional capture technologies are also being developed to allow the equivalent of turning an object around, varying the light source and zooming to whatever level of detail is optimal for the particular questions being investigated. While the ISFDL has focused on providing access to images for scholarly study, educational access needs to be systematically developed as well. Access to transcriptions and possible translations should be developed that do not focus on a single reading of a given text but provide access to the work that results from the foundation of advanced image documentation of ancient texts. Collaborative workspaces should enable teams of scholars world-wide to facilitate coordinated work on the reconstruction and reading of specific texts

---

Fact'. Although InscriptiFact has been actively accessible for just a few years, it has caught on quickly in the field of ancient Near Eastern text studies and is currently actively employed by scholars and students in 40 countries and on every continent except Antarctica.

and collections of texts. New functionalities should allow scholars with resource limitations to utilize the resources of larger institutions through the application of distance learning over the internet. No doubt, many more options will open up to scholars as imaging and databasing technologies continue to advance.

In some respects we are reliving the past as technology propels us into the future. In the ancient past the technological revolution in literacy (first documented in the wedges impressed in clay tablets in Mesopotamia and in the inked hieroglyphs penned on papyrus in Egypt followed by the radical simplification and democratization of writing with the invention of the alphabet in the cultures that developed between these two seminal civilizations) permanently and irrevocably accelerated communication. This revolution allowed us for the first time to preserve for the future our accounts (economic and historical) of the past and, in doing so, they changed the way we think. Perhaps it is presumptuous for us to claim that a revolution of like proportions in literacy is unfolding today—history (probably documented in ways beyond what we can now predict) will be the better judge of that. But what is clear is that the opportunities now becoming available to us through the kinds of technologies discussed above are encouraging us to rewrite, reimage, reimagine, and reconstruct the rules of the game in academic scholarship. As this essay aims to demonstrate, if the tangible literary (and other) remains of the past are to be optimally usable in the present, they will require that we plan for the future in ways that are neither concrete nor abstract but are a creative combination of both.

References

Barkay, Gabriel, Marilyn Lundberg, Andrew Vaughn, and Bruce Zuckerman, 'The amulets from Ketef Hinnom. A new edition and evaluation,' *Bulletin of the American schools of Oriental research* 334 (2004) 41–71.

Darnell, John, F.W. Dobbs-Allsopp, Marilyn Lundberg, P. Kyle McCarter, Jr, and Bruce Zuckerman, *Two early alphabetic inscriptions from the Wadi el-Hol* (Annual of the American schools of Oriental research 59, Part II, 2005).

Hunt, Leta, Marilyn Lundberg and Bruce Zuckerman, 'Non-geographic spatial search in a specialized academic digital library', *International journal of technology, knowledge and society* 4, 2 (2008), 127–135.

——, 'InscriptiFact. A tool for the study of ancient Near Eastern inscriptions. Impact on scholarly practice', *International journal of technology, knowledge and society* 1 (2006), 115–126.

——, 'InscriptiFact. A virtual archive of ancient inscriptions from the Near East', *International journal on digital libraries*, Special Issue on the Digital Museum, 5 (2005), 153–166.

Lim, T.H., H.L. MacQueen, and C.M. Carmichael (eds), *On scrolls, artefacts and intellectual property* (Sheffield: Sheffield University Press, 2001).

Zuckerman, Bruce, 'Bringing the Dead Sea Scrolls back to life. A new evaluation of the photographic and electronic imaging of the Dead Sea Scrolls,' *Dead Sea Discoveries* 3 (1996), 188–189.

——, 'Shading the difference. A perspective on epigraphic perspectives of the Kheleifeh Jar stamp impressions,' *Maarav* 11 (2004), 233–252, 259–274 (illustrations).

Zuckerman, Bruce, Marilyn Lundberg, and Leta Hunt, 'InscriptiFact usability study. Report to the Andrew W. Mellon Foundation' (Unpublished manuscript, 2005).

ANCIENT SCRIBES AND MODERN ENCODINGS: THE DIGITAL CODEX SINAITICUS

David Parker

The Codex Sinaiticus is one of the two oldest Bibles in Greek, and contains the oldest complete New Testament. It is currently housed in four locations (London, Leipzig, Sinai and St Petersburg). Since 2002, it has been the subject of a major project which includes a formal account of its history, a conference, a print facsimile and a popular book, conservation and physical description, fresh digital imaging, a TEI-compliant electronic transcription, translations, and articles of both a specialist and a general nature. Project information and resources are publicly accessible from the Codex Sinaiticus website (www.codexsinaiticus.org). This provides the context for the text that follows. The chapter focuses on the transcription, which was made in Birmingham University, at the Institute for Textual Scholarship and Electronic Editing. It describes the concept and methodology of the project, and compares it with the process by which the manuscript itself was made in the fourth century. In particular, the workings of the team of scribes and the decisions they had to make are outlined, and a comparison is made with the process of making a complex web-based edition. Finally, the significance and likely impact of editions which make hitherto almost inaccessible material available to everyone with a browser is discussed.

Introduction

The Codex Sinaiticus is a manuscript of the Bible in Greek (see Plates 23–25), written in the middle of the fourth century. Although about half of the Old Testament is missing, it includes the oldest known complete version of the New Testament. It contains more books than the standard western Bible, since the Septuagint (the principal ancient Greek translation of the Old Testament) includes writings absent from the Hebrew canon. In addition to the New Testament, the Codex Sinaiticus contains two other ancient Christian writings, the Epistle of Barnabas and the Didache.[1]

---

[1] The best short introduction to the manuscript is S. McKendrick, *In a monastery*

A number of factors single out the Codex Sinaiticus as one of the most remarkable manuscripts ever to have been produced. First of all, it is very large: with a page size of 43 cm high by 38 cm wide (15 inches by 16.9), making it about three-quarters of a metre wide when lying open, it is the largest format Greek Bible extant from any period. Then there is the quality of the parchment, which is remarkably thin and smooth. Next, the number of extant complete Greek Bibles is tiny compared to the total number of codices of the Greek Bible. There are only about fifty in all, of which four date from the fourth and fifth centuries; the rest are much later. Moreover, the number of corrections (over twenty thousand) and their frequency far exceed those found in any other biblical manuscript, and so far as I know any other ancient manuscript. These factors are all the more remarkable when Codex Sinaiticus' place in the history of the book is considered. The codex form itself was only popularised by Christians in the copying of the texts which came to make up the New Testament, and until the fourth century only papyrus had been used in their manufacture. The composition of Christian codices of the third century shows how experimental the codex format still was at that time: examples include single quire codices and a codex consisting of single-sheet gatherings, as well as more normal arrangements. These codices were restricted in size by the strength of papyrus, and could only contain a comparatively small number of writings, such as the letters of Paul, or at the most the Gospels and Acts. Most seem to have consisted of a single Gospel. Only in the fourth century, after the Peace of Constantine and in more settled and prosperous times, did the Christian book begin to acquire its now familiar monumental character. So, at a point when the making of parchment codices was in its infancy, and at a time when the concept of the Bible had hardly come into being, a complex single-volume parchment Bible was produced of a quality which in some respects has never been matched.

To contain these books (48 of the Septuagint, 27 of the New Testament, and two more—77 in total), a codex was required containing between 730 and 740 leaves (1,460–1,480 pages) in 95 quires. The practical challenges of this will be considered shortly. Before that, the history of the Codex must be described if the current project is to be understood.

---

library. For the project see also S. McKendrick and J. Garces, 'The Codex Sinaiticus project: 1. The book and the project', 148–152.

The history of the codex

Where the Codex Sinaiticus was written is unknown, although a place in the Near East is probable. Caesarea has often been considered a possibility. At some point, probably before the end of the first millennium, it came to the world's oldest monastery, St Catherine's on Mount Sinai. There it remained until the nineteenth century. At some point towards the end of this period it suffered some dismemberment: a number of surviving leaves were cut or folded to make book binding materials. In 1844 the German scholar Constantin Tischendorf was shown some of the Old Testament part of the codex, and was given 42 leaves, which he took back home. They are in Leipzig University Library. Shortly after, the Russian Archimandrite Porphyry Uspenski found some of the pieces used in book repair and took them to St Petersburg, where they remain. In 1859, Tischendorf was shown a great deal more of the manuscript, including all of the New Testament and Barnabas. These leaves (347 in total) were subsequently presented by the monastery to the Tsar in 1869. In 1933 they were sold by the Soviet government to the British Museum for £100,000 (part of this sum was raised by public subscription, the first example of this in the UK). In 1975 over one thousand complete and incomplete manuscripts were found in a blocked-off room at St Catherine's, including more of Codex Sinaiticus. These sections are still in the monastery. This is the story in brief of how the manuscript came to be dispersed across four libraries. In sum, the extant portions consists of

43 leaves in Leipzig;
347 leaves in London;
fragments of 5 leaves in St Petersburg;
12 leaves, 8 larger fragments and many tiny fragments in St Catherine's.

Setting up the project

The events surrounding the removal of the majority of the surviving leaves from St Catherine's have led to a great deal of bitterness. Was it legal? Did the monastery approve? Did Tischendorf behave fairly? These have been significant questions which for long went without informed answers. Providing those answers has been one of the main goals of the project to make a virtual Codex Sinaiticus. When on 9 March, 2005 the Archbishop of Sinai, the Chief Executive of the British Library, the Director of Leipzig University Library and the Deputy

Director of the National Library of Russia met to sign the agreement
which underlies the entire project, one of the goals was that they
would produce an agreed account of the events of 1859–69. Underlying the account has been a thorough study of the original documentation, undertaken by several independent researchers. The ability of the
four organisations with custodianship of the manuscript to cooperate
in confronting divisive issues at the same time as managing a complex
project is the most remarkable aspect of the undertaking. What follows
with regard to the concept and execution of the project must be considered within this context, in which an online 'virtual' codex consisting
of new, digital images, a complete transcription and description (some
for the general public, some more specialist) combine with conservation, fresh research, printed material, a conference and a facsimile, in
a modern collaboration which overcomes past difficulties.[2]

   This collaboration extended beyond the four key partners, to include
the Society of Biblical Literature, the Institut für neutestamentliche
Textforschung in Münster, the Institute for Textual Scholarship and
Electronic Editing in Birmingham and other groups and individuals. What follows will concentrate on the role played by Birmingham,
which was to make a complete transcription of the manuscript. The
task was to bring the separate parts—402 leaves and some fragments
in four locations—together in a virtual Codex Sinaiticus.

Resources for research before the project

Some materials were already available to scholars wishing to study the
pages of the manuscript. First of all, we have Tischendorf's remarkable transcription of all the manuscripts known to him, printed in type
specially designed to match the appearance of the manuscript. This
appeared in the 1860s.[3] Secondly, we have the facsimile edition pro-

---

   [2] The first stage of the website (www.sinaiticus.org), containing the books wholly or
partly in Leipzig, the Psalms and the Gospel of Mark, went live on 24 July 2008; the
second, with more of the Old Testament, came out on 22 November of the same year,
and the last on 6 July, 2009, coinciding with the conference and the publication of a
general book (D.C. Parker, *Codex Sinaiticus*). The website includes a detailed page-by-
page conservation assessment, the transcription, selected translations and general and
more specialised articles. The facsimile will appear in 2010.
   [3] The principal edition is C. Tischendorf, *Bibliorum Codex Sinaiticus Petropolitanus*. For full bibliographical details, see C. Böttrich, *Bibliographie Konstantin von
Tischendorf (1815–1874)*.

duced by Kirsopp and Silva Lake in the early twentieth century.[4] Less important (because they are dependent on Tischendorf's edition) are two collations of the New Testament leaves. One, by F.H.A. Scrivener, dates from 1864.[5] In the same year, E.H. Hansell added a collation as an appendix to his edition of the then-known oldest majuscule manuscripts.[6] How he must have wished that either the New Testament leaves had been published a little earlier or he had started work a little later.

For the physical study of the manuscripts, we have been largely dependent on the detailed study written by Milne and Skeat after the acquisition of the leaves from St Petersburg, published in 1938. More recently, two doctoral theses have focused on detailed study of the production of the book and the character of the long series of corrections. Dirk Jongkind's Cambridge research dealt in particular with the way the Old Testament was put together, while Amy Myshrall in Birmingham studied the corrections to the Gospels.[7]

Tischendorf's transcription was magnificent as a resource in the days before photography. The Lake facsimile was excellent in its time. But the one is a printed imitation, the other is in black and white and lacks the detail of a digital image. A comparison with the new images shows how far it falls short of modern standards. Exemplary though Milne and Skeat's researches were, the study of fourth-century manuscripts has advanced in sixty years, while new technology gives opportunities they never had for collecting, examining and comparing information. For example, they had to rely on memory and their notes as they sifted through the thousands of corrections. A researcher can now search the electronic transcription to make a similar analysis.

The concept of digital transcription

Given this situation, the time was undoubtedly ripe for a new digital transcription. The concept underlying it was the electronic transcription

---

[4] H. Lake and K. Lake, *Codex Sinaiticus Petropolitanus*.
[5] F.H.A. Scrivener, *A full collation of the Codex Sinaiticus*. Available on Google Books.
[6] E.H. Hansell (ed.), *Novum Testamentum Graece*.
[7] H.J.M. Milne and T.C. Skeat, *Scribes and correctors of the Codex Sinaiticus*. See also their pamphlet *The Codex Sinaiticus and the Codex Alexandrinus*; D. Jongkind, *Scribal habits of Codex Sinaiticus*; A.C. Myshrall, 'Codex Sinaiticus, its correctors, and the Caesarean text of the Gospels'.

as imagined and developed by Peter Robinson, and subsequently adapted for the Greek Bible by the Institut in Münster and the International Greek New Testament Project.[8] This concept is central to the activity of the Institute for Textual Scholarship and Electronic Editing (ITSEE), which is part of the Department of Theology and Religion at the University of Birmingham. ITSEE is currently involved in projects to make electronic critical editions of the New Testament in Greek, Latin, Coptic, Syriac and other languages; *The Canterbury Tales*; Dante's *Divine Comedy*; Charles Darwin's *On the Origin of Species*, and other texts, as well as developing specialist software and text-critical theory. The foundation of the concept is that witnesses to be included in the edition are transcribed in full. These transcriptions may then be compared using the Collate software to generate a critical apparatus. The edition should present the transcriptions (wherever possible alongside digital images of the witnesses), the apparatus, tools to navigate and search the edition, and descriptive material. Because the apparatus may be exported to a database format, software can be used to study the relationships between the witnesses, using for example the Coherence-Based Genealogical Method or phylogenetic tools.[9] Searching for grammatical forms can be extended beyond the usual printed text environment to include all the witnesses from which a printed text is derived.

While the traditional format of the critical edition has been successful for bringing together editorial materials, it has not been able to include the full transcriptions and images and search tools which the electronic edition provide. However, not even the electronic edition is necessarily able to offer a congenial format for an exhaustive edition even of one manuscript. Nor is it obviously the ideal vehicle for the full range of indexing and paratext that is becoming necessary as mass

---

[8] The Münster products include New Testament Prototypes, the digital Nestle-Aland and the Editio critica maior. Details of all these editions are available at http://www.uni-muenster.de/NTTextforschung/. For the International Greek New Testament Project see http://www.igntp.org/. For Robinson's theories, see P. Robinson, 'What is a critical digital edition?'. For editions made using his software, see www.itsee.bham.ac.uk.

[9] For the Coherence-Based Genealogical Method, see the bibliography at http://www.uni-muenster.de/INTF/; for phylogenetic methods, see for example P. Van Reenen and M. Van Mulken (eds), *Studies in stemmatology*; C.J. Howe et al., 'Manuscript evolution'; P. van Reenen et al. (eds), *Studies in stemmatology II. Kinds of variants*.

digitisation of manuscripts gathers pace.[10] The Virtual Manuscript Room that is being developed in Birmingham and Münster provides a structure for researchers to find and view together the available resources on the web referring to chosen texts and manuscripts (for example, to harvest all the images of witnesses containing the verse 'In the beginning was the Word', along with transcriptions and editions).[11] Work consists of creating software and 'priming the pump' with indexing and other metadata. In time accredited researchers will be able to use the Virtual Manuscript Room as a sort of wiki, adding metadata to manuscript entries.

There are thus three ways of presenting material according to the Birmingham and Münster model: in the Virtual Manuscript Room, as a mash of what is available; in the electronic edition, which combines traditional philology with transcriptions, images and search tools; and in what one might call the de luxe edition of certain manuscripts.[12] It is to the third category that the transcription of Codex Sinaiticus belongs.

Making the transcription

The transcriptions were made in plain text format, using pre-defined tags. They were then exported into tei-compliant XML and delivered to the website developers in Leipzig for final processing in HTML. Accuracy was achieved by making two separate transcriptions. Each was produced by adapting an existing electronic text (except for Barnabas, which was keyed in *de novo*). These were then automatically compared using the Collate program, and all differences were reconciled against the images. An important aspect of the project was that all images should be seen by the same researchers, who also examined

---

[10] Ten years ago a typical project was the Early Manuscripts at Oxford University website (http://image.ox.ac.uk/list?collection=all), containing images of ninety manuscripts from seven Oxford libraries. Two years ago a typical project was the Codices Electronici Sangallenses (http://www.cesg.unifr.ch/en/index.htm). The goal is to make all the manuscripts in this famous library available online; with the latest update on 23 December, 2008, the total available reached 251. A typical project today is e-codices, providing access to images of all the medieval codices in Switzerland.

[11] Work in Münster is concentrating on manuscripts of the Greek New Testament, and in Birmingham on the University's Mingana Collection, one of the world's largest holdings of Christian oriental manuscripts.

[12] The British Library's edition of the single manuscript of *Beowulf* was one of the first examples of such an edition.

the manuscript itself in all locations. These researchers were Amy Myshrall and Tim Brown, who made all the transcriptions. They visited Leipzig twice, and St Petersburg and St Catherine's once each. Rachel Kevern was responsible for initial reconciliation of transcriptions, and she and Hugh Houghton undertook XML conversion, checking and proof-reading.

The web-based edition contains one new element with several important uses. The images and transcription are aligned so that if the mouse is pointed at a word in either, it is highlighted in the other. This is useful for people who do not know very much about Greek manuscripts, and can function as a tool for self-tuition in Greek palaeography. It is also useful for more expert users, especially in reading difficult pages where the ink is faint, as a tool for orientation and for monitoring the transcription. Beyond this, the feature places the images, and thus the manuscript, at the heart of the presentation, the transcription functioning as interpretation of the manuscript rather than as its surrogate.

This feature therefore singles out the transcription of Codex Sinaiticus from others we have so far produced of the New Testament. Those function within the context of other transcriptions and a critical apparatus.[13] Here, our goal is not a transcription to operate as one of a number of starting-points for making a critical edition. Rather, the goal is found within the manuscript itself, the material entity which, even though it is incomplete and divided between four locations and is one among thousands of copies of these texts, deserves attention as a unique object. Behind this lies the concept of the biblical text as available to us only as a set of witnesses, each representing the text as it was reproduced at a particular time and place. At no point does the significance of these individual entities become inferior to an editorially reconstructed archetypal or 'original' text.

It should be clear by now that this has been a very complicated project to manage, so far as coordinating teams with different responsibilities in different places is concerned. This management was undertaken by the British Library, who appointed a Project Manager (Dr Claire Breay) and Curator (Dr Juan Garces), under the direction of Dr Scot McKendrick, the Head of Western Manuscripts. Budgets, coordination of working parties and of many different phases and spe-

---

[13] See for example the editions available at www.iohannes.com.

cialisations and time schedules have been among their responsibilities, along with the core role of working with the other three libraries. It is arguable that these organisational processes have been as significant for scholarship as the result, involving as they have such extensive collaboration, as well as the setting of new standards in conservation, documentation, digitisation, transcription and other areas. One of the most striking facts of all is that there is a very strong parallel between the demands of creating the digital Codex Sinaiticus and the making of the manuscript itself back in the fourth century. Exploring the task facing the scribes and comparing it with today's counterpart casts light on both.

Ancient practices mirrored today

An analogy

It has already been noted that the parchment codex and the complete Bible were novelties when Codex Sinaiticus was made, just as the electronic edition is today.[14] The virtual Codex Sinaiticus Project has brought to light similarities between the ancient and modern teams. In fact the justification for the following account of the manuscript's production is that the procedures followed by the ancient team can be viewed as a model for the scholars of today, while the terminology which we use for our project may be used to describe the way in which the scribes worked. The people who made the manuscript will have had very few precedents, and must often have had to make their own decisions, sometimes as they went along. What do we know about the process, and what can we reconstruct of it, or even guess about it?

Making the Codex Sinaiticus

First of all we have the production team. We know of three scribes, A, B and D. Of these, D was plausibly the senior, since (1) his leaves are the best and most accurately written and (2) he sometimes wrote replacement leaves in sections by his colleagues, and added ancillary material. We now know that he wrote at least some of Genesis, as well as the first part of the Psalms, and these are the longest blocks

---

[14] For the field of New Testament manuscripts and textual scholarship, see D.C. Parker, *An introduction to the New Testament manuscripts and their texts*.

that he contributed to the manuscript as it now survives. Since the former is at the beginning of the codex, and the latter is the first of the poetical books, which are written with two rather than four columns to the page, it is possible that each time he was providing a template for his colleagues. Scribe A wrote the highest proportion of the extant leaves. He was a competent scribe who suffered only from a tendency to omit text (a common enough failing among copyists who were paid by pages of exemplar copied and not by the hour). Scribe B, who wrote the smallest proportion of what survives, was the poorest of the three, showing a strong tendency to write the text as he pronounced it rather than according to the roles of orthography.

What lay behind the three scribes? A wealthy private patron? An institution? What were they producing? A de luxe copy to satisfy a bibliophile? A grand copy for a church? Who else was in their workshop? Did they have other people to do some of the rougher work, such as preparing the vellum and keeping them supplied with raw materials? The connection between the parchment manufacture and the production team must at any rate have been close. It is worth noting that a good many animal skins (both sheep and bovine) went into this manuscript. Although most skins supplied two sheets, some single sheets were taken from smaller skins.

To make a complete Bible, this team had first to collect together a set of codices containing the separate parts. For the New Testament, this must have consisted of at least five, possibly seven, or even ten separate codices, depending on whether they possessed the Gospels and Acts and the Catholic Letters in separate volumes. For the Old Testament, if the contents of the surviving papyri are anything to go by, they may have needed to collect in the order of twenty-four different manuscripts.

These codices, probably over thirty in number, are likely to have been copied by as many different scribes, were written on papyrus or parchment, in different scripts, with different layouts, and showing an assortment of orthography, punctuation habits, quality, and other aspects of writing. The textual and transmission history of each of the different parts will have been unique, and of course the range of literary genres caused another problem. The Sinaiticus team had to make a series of decisions which would bring all these things into a consistent whole.

First of all, they had to agree on technical standards, starting with the quality and preparation of the parchment. The technology needed to produce thin parchment was essential to the whole operation, since it is only because the pages are so fine that the enterprise was possible in the first place: a complete Bible on thick parchment would have been too fat to be a single bound volume. The page size and page layout must have had to be determined before the first parchment sheet was prepared, perhaps even before suitable skins (still on the animals' backs?) were selected. To determine this the scribes must have planned the number of lines and columns, and so of the letter size. It follows that an agreed script must have been an early decision. An agreed formula for preparation of the ink (both black and red) was also necessary, and one wonders how this was carried out—was a single person responsible?

The scribes must have also made a number of early decisions about the order of the books. Starting with Genesis and the other books of Moses may have been an obvious choice, but the oldest Bibles do not have the same order of biblical books in either testament. Here they may have kept their options open for as long a possible. The codex is so designed that each major book or block begins on a new gathering. This was achieved, unless it worked out right by good fortune and a little compression or expansion of the text density, by adjusting the number of leaves in a gathering. For example, the last two sheets of Gathering 58 were cut out, so that Malachi (the last of the prophetic books) ends on Quire 58, Folio 6v and the Psalms begin on Quire 59, Folio 1r. By matching the number of leaves in a gathering to the size of individual texts, the scribes were free to copy the books in any order and then bind them in the order that was eventually chosen.

Once they got to the stage of having parchment ready for writing, they had then to develop a system to prick and rule the leaves, preferably with an economical workflow, taking account of the need for running titles, the different lay out of the poetical books (two columns per page instead of four), paragraph numbering in the Gospels and indentation. There had also to be developed a typical wording for the running titles, titles and subscriptions of books. The last of these included decorative patterns (coronae), which are very varied in design. It is worth noting this, because it shows that there were matters on which the scribes must have agreed not to standardise their practice. Trying to achieve unworkable consistency is perhaps as damaging to a successful project as forgetting to agree on what has to be done in a single way. Either our scribes worked this out, or they simply forgot to discuss, or failed to agree, this feature (or ignored their decision).

Even now, they are not in a position to start. They have to work out the division of labour. This required them to calculate the number of quires needed for each block of material. This is where their use of a number of exemplars, each with its own lay out, will have posed a problem, since the conversion of each into the new lay out of their codex will have required a different calculation. It may be that, like a typesetter, scribes acquired skill in recalculating a lay out which meant that they did so easily. But, without going into detail, one can sometimes observe a rather complicated process which suggests that they set out with an allocation of material which was then modified, for example in the way that scribes A and D alternate in 1 and 4 Maccabees, at the same time struggling to fit their material into its allotted quires.

The challenge was formidable. How difficult the whole process really was for the scribes can be seen from various details. For example, they managed to duplicate fourteen pages of the Old Testament and never noticed; they omitted 2 and 3 Maccabees completely (was this connected with the alternation between the scribes in these books, because each assumed the other was copying them?);[15] they miscalculated the space needed in the transition from Revelation to Barnabas; they seem to have had some kind of difficulty with providing the Psalm numbers and titles.[16]

Once the manuscript was written (I gloss over the amount of work required), a good scriptorium would check it. The scribes of Codex Sinaiticus carried out this task carefully. Admittedly, they seem to have overlooked the repetition of fourteen pages, and to a modern proof reader or editor they missed more than they corrected. Nevertheless, they made over 2,200 corrections, a formidable number by the standards of the age.[17]

---

[15] Curiously, the team producing Codex Vaticanus (another Greek Bible written in the middle of the fourth century) made a very similar mistake, omitting all the books of the Maccabees.

[16] Dirk Jongkind has drawn attention to the difficulties of some of the transitions. Between Judith and 1 Maccabees, the evidence indicates that Scribe A wrote 1 Maccabees on Quire 39, Folio 3r onwards before D had written the end of Judith on the first two sheets. This indicates amazing confidence, even though D had to space out the end of Judith and they had to leave one column blank.

[17] A search of the XML gives 2,225 occurrences of the tag 'S1', indicating a scriptorium corrector. Since not all corrections can be ascribed so closely, the true number must be higher.

There is an advantage and a disadvantage to being a pioneer. The advantage is that there is no weight of expectation. The disadvantage is that you are laying the groundwork for somebody else to do the same thing better. The makers of Codex Sinaiticus are remarkable in that, imperfect though their work is, they produced something which has never been matched. Their impressive organisational achievement meant that they created a book which looks remarkably homogeneous—and above all is an artefact of great beauty.

It is worth adding that there is some evidence of another team at work, in a very detailed process of correction carried out a couple of centuries after the manuscript had been produced. Known as the 'c' group of correctors, and divided into ca, cb (1, 2 and 3), cc, cc* and cpamph, together these revisers were responsible for about 22,000 corrections (some are corrections of corrections), an average of about twenty-nine per page. The majority are in the Old Testament, peaking at forty-seven to a page in the Minor Prophets.[18] It is not certain how closely these correctors worked together. But the fact that such a rate of revising is unprecedented suggests a connection between them. If this is indeed the case, then the virtual codex is the product of the third team, this time responsible for the text as it appears on the screen.

I leave it to the reader to draw out the parallels between the fourth-century scribes and today's virtual Codex Sinaiticus in more detail. But of course there are differences. The biggest is that the modern project has not interfered with the text on the page (except to the extent that the conservators have stabilised fragile areas of the parchment).

Transcription as interpretation

The focus has been on recording and interpreting what is written. In a moment of enthusiasm, I once expressed the goal of recording every ink mark on the page. Actually, this was not practicable. Should every stray blot and smudge be noted? The outlay of time and therefore money probably does not justify it. Should the transcription also play a role in interpretation of the ink on the page? It cannot fail to do so, since for example some accidental marks may be indistinguishable from punctuation marks. The act of making a transcription includes

---

[18] The details (as in February, 2009) are: ca 14584; cb1 500; cb2 1025; cb3 4620; cc 1001; cc* 27; cpamph 369.

many interpretative decisions, and it is appropriate that those who have spent two years making the transcription should be trusted as reliable guides.[19]

Nevertheless, there are boundaries to the degree of interpretation that is appropriate, and we deliberately stepped back from paying too much notice to new features and research questions that emerged in the course of the transcription. Foremost is the possibility that Scribe B should be divided into two scribes (on the grounds of possible differences in script). Second is the discovery of some 'mystery marks', short oblique strokes through letters in certain places. So far there is no certain explanation for them. Are they pen strokes made when counting letters (for some unknown reason)? Are they caused by debris on the pen? Why do they seem to occur randomly and infrequently, with a number of them occurring together and then none for pages? The process of tagging also drew to our attention a number of questions relating to punctuation of various kinds. There are different shapes of *paragraphoi* (horizontal strokes dividing sections of text); do these differences indicate slightly different uses, or are they different versions of the same symbol?[20] Sometimes we find punctuation in the form of three or even four points; what might one learn from them? All these are interesting research questions. But their proper study has had to wait.

What is the purpose of this electronic edition? If there is a single answer, it concerns access. Until this project, study of the manuscript had been hampered by its geographic dispersion to four locations and the variable conditions of access and preservation. Tischendorf's editions and the facsimile are themselves rare books, and lack the New Finds of 1975, which have hitherto remained unpublished; access to the leaves themselves is restricted in order to conserve them, and to see them all would in any case be a formidable logistical undertaking. The virtual edition makes available to every scholar images which are easier to read than the manuscript itself, a fully tagged transcription and the page-by-page description of the conservation document.

---

[19] Dr Myshrall, having worked with the codex for many years, is able to recognise the scribes and correctors through detailed knowledge of the characteristics of each.

[20] The distinctive shape of some symbols defeated the current range of characters in Unicode, so that we displayed an image of a selected representative of that symbol at each of its occurrences.

Thus access for experts is assured. It may prove even more significant that the virtual edition also makes the manuscript available to everybody with a browser, wherever they may be. Specialist textual research has always been restricted by access to the primary materials in a few libraries (the situation was no different in the fourth century and after, and one intriguing aspect of Codex Sinaiticus and its correctors is that they may have connections with the famous library of Caesarea). Today, that restriction is removed. The future is fascinating: how will this unprecedented level of access affect textual scholarship? How will the role of scholars change? What attitudes will new groups of users have to texts so freely available to them? How and to what extent will a wider public engage with these new materials? Will we regain the textual richness of the times before standard printed editions? Finding out some of the answers to some of these questions is going to be one of the most exciting aspects of textual research in the next generation.

Conclusion

Electronic editions of manuscripts such as the Codex Sinaiticus are always going to be a rarity. One could not justify the outlay of resources except in the case of the most precious treasures. This edition presents the very limits of what is currently possible in a web edition. It thereby does justice to the Codex Sinaiticus, which in its day, and still, represents the summit of the art of reproducing a complex text in manuscript form. The words written by one of the correctors of the manuscript in describing the copy against which he had made his corrections apply also to the Codex Sinaiticus and, we hope, to our reproduction of it:

> If it be not presumptuous so to say, it would not be easy to find a copy equal to this copy.[21]

---

[21] From the colophon to Esther, translated by T.C. Skeat, 'The use of dictation in ancient book production', 3–32, 18.

References

Böttrich, C., *Bibliographie Konstantin von Tischendorf (1815–1874)* (Leipzig: Leipziger Universitätsverlag, 1999).

Hansell, E.H. (ed.), *Novum Testamentum Graece, antiquissimorum codicum textus in ordine parallelo dispositi accedit collation codicis sinaitici*, 3 vols (Oxford, 1864).

Howe, C.J., A. Barbrook, B. Bordalejo, L. Mooney, P. Robinson and M. Spencer, 'Manuscript evolution', *Trends in genetics* 17, 3 (2001), 147–152.

Jongkind, D., *Scribal habits of Codex Sinaiticus* (Texts and Studies, third series 5) (Piscataway: Gorgias, 2007).

Lake, H., and K. Lake, *Codex Sinaiticus Petropolitanus*, 2 vols. (Oxford: Clarendon Press, 1911 and 1922).

McKendrick, S., *In a monastery library. Preserving Codex Sinaiticus and the Greek written heritage* (London: The British Library, 2006); German translation, ed. U.J. Schneider as *Codex Sinaiticus. Geschichte und Erschliessung der Sinai-Bibel* (Leipzig: Universitätsbibliothek, 2006).

McKendrick, S. and J. Garces, 'The Codex Sinaiticus Project: 1. The book and the project', in *Care and conservation of manuscripts 10. Proceedings of the Tenth International Seminar held at the University of Copenhagen 19th–20th October 2006*, eds G. Fellows-Jensen and P. Springborg (Copenhagen: Museum Tusculanum Press, 2008).

Milne, H.J.M. and T.C. Skeat, *The Codex Sinaiticus and the Codex Alexandrinus* (2nd edn, London: British Museum, 1955; 1st edn 1938).

——, *Scribes and correctors of the Codex Sinaiticus, including contributions by Douglas Cockerell* (London: British Museum, 1938).

Myshrall, A.C., Codex Sinaiticus, its correctors, and the caesarean text of the Gospels (unpublished Ph.D. thesis, University of Birmingham, 2005).

Reenen, P. van, A. den Hollander and M. van Mulken (eds), *Studies in stemmatology II. Kinds of variants* (Amsterdam: Benjamins, 2004); online version at http://site.ebrary.com/pub/benjamins/Doc?isbn=1588115356.

Reenen, P. van and M. Van Mulken (eds), *Studies in stemmatology* (Amsterdam: Benjamins, 1996).

Parker, D.C., *Codex Sinaiticus. The story of the world's oldest Bible* (London: The British Library and Peabody, Mass: Hendrickson, 2009).

——, *An introduction to the New Testament manuscripts and their texts* (Cambridge: Cambridge University Press, 2008).

Robinson, P., 'What is a critical digital edition?', *Variants* 1 (2003), 43–62.

Scrivener, F.H.A., *A full collation of the Codex Sinaiticus with the received text of the New Testament, to which is prefixed a critical introduction* (Cambridge: Deighton, Bell and Co. and London: Bell and Daldy, 1864; 2nd edn, 1867).

Skeat, T.C., 'The use of dictation in ancient book production', *Proceedings of the British Academy* 42 (1956), 179–208, repr. in *The collected Biblical writings of T.C. Skeat*, ed. J.K. Elliott (NovTSuppl. 113; Leiden and Boston: Brill, 2004).

Tischendorf, C., *Bibliorum Codex Sinaiticus Petropolitanus*, 4 vols. (Leipzig, 1862; repr. Hildesheim: Olms, 1969).

TRANSMITTING THE NEW TESTAMENT ONLINE

Ulrich Schmid

The Greek New Testament poses great challenges to any attempt to provide a scholarly critical edition for at least two technical reasons. Firstly, we have a huge number of manuscript witnesses to digest—at this moment we know of no less than 5555 entries in the official list of Greek New Testament manuscripts. Secondly, we have only tiny proportions of the manuscripts from the first millenium preserved, which forces incorporation of early translations into Syriac, Coptic and Latin into an edition of the Greek New Testament in order to partially compensate the losses.

This chapter describes and illustrates the traditional workflow that has to be maintained in order to produce such a critical edition in print. It also attempts to sketch a digital workflow that integrates images of the manuscripts with transcription (incl. manuscript studies and cataloguing), apparatus building, stemmatological analysis and publishing. Such a digital workflow should be designed in a way that makes full use of the advantages of the electronic medium (collaborative, modular, updatable) while at the same time meeting the accepted standards of the printed book (sustainability, accessability, accountability). It goes without saying that producing scholarly editions along theses lines involves new perspectives and challenges on, inter alia, issues of copyright both on the level of institutions that own the artefacts (manuscripts) and on the level of collaboration with traditional print publishers.

For almost 2000 years the Christian Bible has participated in the history of the transmission of text. Moreover, it is so intrinsically and prominently linked to it, that for large parts of its history the Bible can be viewed as a synonym for the book.[1] In order to illustrate this bold claim, let me draw your attention to two major media revolutions in the history of the book. The first is the transition from scroll

---

[1] As Christopher de Hamel writes in his introduction to *The Book. A history of the Bible*, 'The title...is *The Book. A history of the Bible*, but it could as well be *The Bible. A history of the Book*'.

to codex as the dominant medium for the publication of literary texts. This transition started some time in the first century of the common era and the New Testament appears to be the first book from antiquity to have been published in the codex format from the very early stages onwards.[2] The second media revolution is the transition from the manuscript copy to the printed book, which started in Europe in the middle of the fifteenth century initiated by Johannes Gensfleisch, also known as Johannes Gutenberg from Mainz. The most prominent object from his workshop is the 42-line Gutenberg Bible.[3] Now what are the consequences of the digital media revolution of the late twentieth century for the transmission of the New Testament? Or to put it differently: What are the specific challenges modern editors of the New Testament face? And how can the digital medium help us to meet these challenges—perhaps even more effectively than previous techniques?

In addressing these questions I intend to sketch what I would call a fully integrated digital edition ('fide') of the New Testament that will ultimately be produced online as well. I should emphasize right from the beginning, that I am concerned with scholarly editions only. What I have in view is an edition for research and teaching purposes. My contribution develops as follows. First of all the focus is on the traditional ways of editing the New Testament, which I call 'analogue'. Secondly, I shall be sketching a digital workflow that aims at a fully integrated digital edition. Finally, I want to touch upon some issues and obstacles that need to be looked at when pursuing the digital way of editing.

Editing the New Testament in an analogue manner

Making an edition involves three steps:

(a) collecting data, i.e. manuscript(s) and information from the manuscript or manuscripts that are supposed to be edited have to be retrieved and assembled;

(b) processing data, i.e. the information that is collected has to be selected and organized in order to be made intelligible and meaningful;

(c) displaying data, i.e. the grouped and meaningful information finally has to be output in order to be accessible to potential users.

---

[2] Harry Y. Gamble, *Books and readers in the Early Church.*
[3] De Hamel, *The Book*, 190–215.

These three steps are usually accomplished by breaking up the work into a series of operations that logically and pragmatically build upon each other. An excellent account of the resulting workflow for classical texts has been given by Martin L. West.[4] In what follows I intend to present editorial workflow examples from the realm of New Testament editions in order to appreciate the challenges that modern editors of the New Testament face. Let us start with a famous seventeenth-century example, the Walton Polyglot.[5] At the top left side of an opening one finds usually about ten verses of Greek text with interlinear Latin translation. The apparatus that follows below that is a tiny sample of variant readings from one manuscript, covering just a single line. The rest of the opening is designed to present ancient translations of the same ten verses: the Latin Vulgate, the Syriac, the Persian, the Arabic, the Ethiopic etc. The editor in this case opted for a convenient display of the wealth of ancient versional evidence. Here we find the first challenge for editors of the New Testament, namely the multilingual tradition of the New Testament that in part harks back to the second century CE, thus potentially representing very early textual data that might be lacking from the Greek tradition available to us today.

Another, more recent example is the International Greek New Testament Project's edition of Luke's Gospel.[6] Here the variant readings from the Greek manuscript tradition dominate the scene. For only one verse from Luke's Gospel we find about twenty lines of apparatus in two columns. The second challenge is therefore that there is not only a wealth of ancient versions of the New Testament, but a host of Greek manuscripts as well. In fact we are talking about a total of more than 5500 items in a list of Greek NT manuscripts. Compared to the Walton Polyglot the ancient versions have lost their prominent position: Arabic, Syriac and Ethiopic are absent. However, the evidence from these versions is still there, but digested within the apparatus to the Greek text. That is to say, the information from the versions has been extracted, for example from the Old Latin, condensed and projected onto the Greek text. And this procedure hints

---

[4] Martin L. West, *Textual criticism and editorial technique applicable to Greek and Latin texts*, 61–103.

[5] An image can be found on http://itsee.bham.ac.uk/parker/introduction/index.html under no. 37a.

[6] The New Testament in Greek. The Gospel according St. Luke. Edited by the American and British Committees of the International Greek New Testament Project, 1984–87.

at the third major challenge for editing the New Testament. This challenge I would call the de-contextualization of the evidence. In Walton's Polyglot one could compare the Syriac as an integral consecutive text. In the IGNTP's edition, however, a Syriac witness has been dissolved into a series of Greek virtualisations of the actual Syriac text. This is a one-way ticket. In case somebody wants to go back to the Syriac to check and verify the context, he or she has to get the relevant information from another publication. This way of representing versional material in an edition of the New Testament is a good illustration of de-contextualisation of the evidence. But that not only applies for versional material. Even for the Greek manuscript evidence we lack the proper context of the readings that are presented in the apparatus. Sometimes it would be helpful to know about the layout of a particular manuscript that is said to exhibit a peculiar reading. How has the scribe segmented and spaced out the text on the page? Are there other phenomena, like commentaries or marginalia interfering with the main text? So in a way, a textual apparatus, even for manuscript witnesses in one and the same language, is by definition a condensed, abridged, truncated representation of the evidence.

Let us look at some more details of the actual workflow that is behind such an analogue edition of the New Testament in order to identify more challenges. How is all the evidence displayed by, for example, the IGNTP's edition of Luke's Gospel actually gathered, processed and arranged? In Figure 1 we find a sheet from a collation done in pencil on paper. It is the collation of a lectionary manuscript. Collations are done by reading a manuscript against a base text while taking down all the differences. This is a standardized procedure. When one spots a difference, the chapter and verse number, the lemma, that is to say the passage in the base text that is differently rendered in the manuscript under scrutiny, and finally the different wording of this manuscript are noted. Any educated user of such a collation can then take a copy of the base text and reconstruct the wording of this manuscript by substituting the indicated words at all the noted points of variation.

However, in the case of our example here such a reconstruction would not result in a true representation of the manuscript. Crucial information on the layout and additional material is not to hand. Even the text cannot be reconstructed completely, because the editors of the IGNTP decided early on that abbreviations should not be collated.

Although this procedure is a very effective way of reducing the volume of information drawn from a manuscript, the passage under dis-

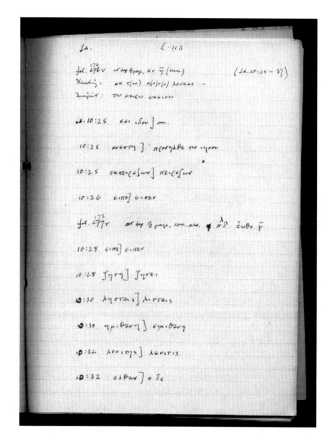

Figure 1: Collation sheet of a lectionary manuscript, IGNTP edition of Luke's Gospel.

cussion is only seven verses from a single witness. It is produced in pencil on paper and the sheet has been made part of a booklet. What would collations of about 250 Greek manuscripts plus other witnesses for about 1150 verses pile up to? But the real challenge is yet to come. How should we imagine processing that amount of data? How would one organize the evidence collected from the many witnesses? The first step could be to prepare sheets, which give the range of known variants at a given verse with huge spacing, so that they can be filled in with the information from the witnesses. Subsequently, masses of collations need to be ploughed through in order to place the sigla of the witnesses to the appropriate readings. Figure 2 displays two sheets of an IGNTP apparatus in the making. Here one can see such a pre-prepared apparatus sheet used to combine the evidence for Luke 2:15.

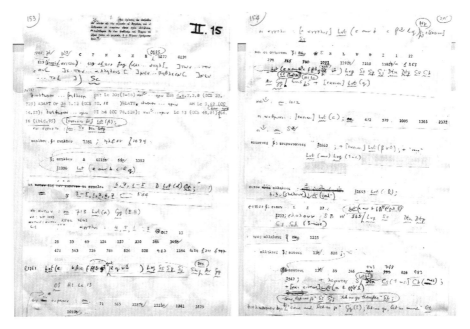

Figure 2: Evidence for Luke 2:15 prepared on a sheet for the IGNTP apparatus

The range of known variants is supplied in typewriting. However, dealing with huge numbers of additional witnesses will inevitably unearth new readings that could not have been thought of in advance. These, then, would have to be added by hand. This is not a problem in the case of a single word. It is a nightmare though when one discovers half way down the road that the lemma reading which has been used up to then does not really fit the complex evidence displayed by the witnesses. Initially a longer printed lemma of seven words was used for quite some time, but then the decision was made to split it up into three separate lemmas, which have been added in hand writing. As a consequence, the information that had been placed with the old lemma had to be redistributed over the new lemmas. Sticky tapes were glued over the old notations, a piece of paper with printed numbers was glued to the page, apparently to revise older information, and additional handwritten notations and corrections can be detected. This is a heavy version of cut and paste, it seems. And there can be little doubt that such working routines are prone to introduce new errors of all kind: missing or wrongly placed information due to oversight, cramped handwriting, misplacement of sigla and the like. The next step was to take this type of apparatus manuscript, containing hundreds of sheets, to the printer, which involved even more work. Of course, it is easy to blame editors for the mistakes their products display, but the pure logistics of producing a textual apparatus of that complexity are such that it is basically impossible to keep it error free. As one can see, the information drawn from the many manuscripts and digested into an apparatus travels a lot. Manuscript readings become divorced from their physical substrate by way of collation against a base text. The readings then get reshuffled in combination with the multitude of readings. And the reshuffling has to be taken literally, as the example from the IGNTP's apparatus in the making shows. Readings and sigla travel on the apparatus page from one place to another, they are erased, cut, pasted, and in many cases it is virtually impossible in retrospect to fully trace their moves from one place to another.

Hence, this is another huge challenge to editing the New Testament in an analogue manner: the interrupted and often obscured chain of moves to which the information is subjected from the stage of data collection to the final product on the printed page. Despite all the admirable efforts undertaken to cope with the logistics of the edition, in essence we face a serious lack of traceability of the data in such an analogue setting. And in addition to that, the huge amount of information that goes into an analogue printed apparatus has been extracted, truncated, and re-shuffled precisely to that effect, i.e. to fit the printed book, to reside between two covers on the book-shelves of our studies or preferably on our desks to be used and studied. But as soon as scholars use and study our New Testament edition, they find all sorts of extra bits and pieces of interesting information, corrections, additional data from manuscripts, versions or Church Father writings, apocryphal texts, etcetera, not used in the edition. And the only way to deal with that is to scribble into the margins of the printed edition. In so doing we testify to the cul-de-sac experience with analogue apparatuses. They are hardly updatable. The only way to remove errors, expand information and customize the availability of the data is by manually re-shaping the apparatus within the printed book. As we all know there are serious limitations to such ways of personal updating. We soon face spatial constraints in the margins and run out of colors to highlight specific information in the edition. Moreover, we can hardly share the additional information with other users, because it is too much work to export our customized data or import data from other users. As a result we have created for ourselves highly specialized and customized editions that are of service to us individually, but are of very limited use to others. This, by the way, is likely the main reason why we are so tied to the edition we started to use early on, even if newer and better editions become available in the course of time. Newer editions may contain fewer errors and more information, but they usually lack not only the look and feel of the old edition, but more importantly the customized personal updates and expansions that we have added to our main working copy over the years. We do not want to scribble all that again into a brand new book.

There is no need to belabor the point. It has become clear that the analogue printed scholarly New Testament edition faces one main challenge, namely the unwieldiness of the sheer amount of data: Greek manuscripts, versional and Patristic data. The efforts to cope with this challenge result in an apparatus with the following major drawbacks: de-contextualization of the evidence, lack of traceability, non-updateability.

Editing the New Testament digitally

Let us now move on to a way of editing the New Testament digitally that wants to make use of the benefits of the electronic medium. As I see it, there are three main benefits that spin off a number of other benefits that are highly significant for both the workflow employed in creating an edition and the way editions are perceived and will be consumed in the future:

1. The first main benefit of the electronic medium is the possibility to establish connections between digital objects. Every digital object exists in bits and can thus be connected to any other digital object. Although our experience as users of personal computers is such that we often encounter incompatibilities between file formats and operating systems, the underlying basic structure of digital information is the same. Incompatibility is due to the differing architectures that are built on top of it and the lack of effective conversion routines between them ('not supported' conversions, as the jargon has it). Annoying user experiences notwithstanding, the basic fact remains that digital objects are in principle connectable.
2. The second main benefit of the electronic medium is the ease with which digital objects can be replicated. 'Copy and paste' and 'save

as…' are among the most frequently used operations in our daily use of personal computers. Although we occasionally experience data loss, this is usually not due to the fact that we could not have prevented this from happening. The mere mechanics of replicating digital objects is emphatically a non-issue in that regard. It is rather a matter of laziness and lack of an appropriate backup routine. But replication is not just about backup, it is also about distribution and delivering. Digital objects can be morphed and converted into formats that can be delivered in hard copy or over the internet to any number of users simultaneously, viewed on screen and printed locally.

3. The third main benefit of the electronic medium is the huge amount of space it provides for digital objects. This is a comparatively recent development, for in the early days of the electronic age the ability to store data was rather limited. Currently, data storage space seems to abound as we can experience on various levels. Anyone who subscribes to Google or Yahoo or similar such services for email, blog or homepage receives ever more server space at ever lower costs. Universities and affiliated libraries provide email accounts and a growing number of additional services to all their users (staff, students and beyond). This is supported by the availability of massive amounts of server space at very reasonable costs. Internal and external hard drives double their capacity every year for the same price.

There are many spin-offs on different levels of these main benefits for creating digital editions. On the level of the workflow we can remodel the series of operations as described in the previous section, thus minimizing the human factor that is likely to introduce new errors and increase transparency and traceability.

The workflow ought to start with a faithful diplomatic electronic transcription of every manuscript selected for the edition. The benefit of the electronic transcription is the ability to mirror the manuscript line by line, column by column, page by page, enabling the lay-out to be replicated and sophisticated ways of encoding corrections, abbreviations, decorated initials to be included, down to the level of punctuation and spacing. That way in addition to the manuscripts' 'readings' we can include a fair amount of non-textual features in a machine readable format, which can be used to support the analysis of textual phenomena surrounding a given reading or be systematically analyzed on their own merits. Especially designed programs enable interactive stripping of any part of the evidence from the manuscript transcriptions that is deemed not useful for displaying in an apparatus without affecting the encoded and stored transcription data. Moreover, transcriptions of many manuscripts can then be processed and automatically collated in order to produce a traditional apparatus. Interactive tools allow for tailoring the variation units to the design an editor or user finds most useful. Stripping data and modelling the apparatus will not affect the integrity of the manuscript information as recorded in the transcriptions and is entirely reversible. Conceived like that the workflow from the collection of data to the display of part of the data as variant readings in an apparatus is entirely transparent.

On the level of project management, the digital medium allows breaking down huge projects into manageable portions, because a modular set-up can be applied. Provided the basic data scheme is sound and appropriate, one can start with just, say, twenty manuscripts. Any number of further manuscripts can then be incorporated later in the previous data set that has been digitally stored. Hence, even newly discovered manuscripts or improved information can be comparatively easily added to update the project. Moreover, a modular set-up enables different individuals or projects to devote themselves to specific parts of the overall project. To give an example: for a digital edition of the New Testament it is entirely conceivable to assign the Latin or Syriac versions to different groups of specialists. As long as each group follows the same data scheme we can work simultaneously in Germany on the Greek manuscripts, in the United Kingdom on the Latin material and in the Netherlands on the Syriac and still exchange our data and feed them into a larger structure that encompasses them all.

Arguably the most important but certainly the most visible impact of the electronic medium on the digital edition takes place when it comes to displaying the data. The fairly straightforward and familiar design of the printed edition with a base text and one or more apparatuses, depending on the complexity of the evidence presented, can be easily reproduced and displayed on screen. But, of course, there is no need to stick to a single design. The standard version of the apparatus that displays variant readings in relation to a word or a number of words in the base text, hence word or lemma apparatus, can be complemented with an apparatus based on larger sense units (sentences, stanzas, or—in the case of biblical texts—verses). Such a verse apparatus could display different versions of a verse aligning with each other, having the differences highlighted by color codes.[7] There are many more options conceivable, such as customized apparatuses based on a limited range of witnesses selected by provenances, dates, or types of manuscripts (papyri, parchment, paper), etcetera. The issue is not only *how* to display the data, but also *what* data should be included in the edition. As soon as we have new data collected by means of electronic diplomatic transcriptions, we can include those as well in our edition and display them—only a mouse click away—in another window alongside with the apparatus. Since the data in the apparatus are taken from the electronic transcriptions the two are deeply interconnected. Hence, it is easy to switch back and forth between a particular variant from a manuscript as displayed in the apparatus and the specific passage of that manuscript's transcription in order to view and study the variant within its immediate unique context.

But the question of what to include does not stop here. The ultimate source for the evidence presented in any edition is/are the object/s itself/themselves. And the presence of the objects within a digital edition is not only conceivable by means of electronic transcriptions but through digital images of the manuscript pages as well. Since digital objects can be linked to each other, it is rather straightforward to display every image of a manuscript page with its respective electronic transcription or *vice versa*. The splendid Codex Sinaiticus Project does exactly that[8] for one prominent biblical manuscript, and the digital edition simply extends this to the selection of manuscripts that are included in the edition. Just as any user of the Codex Sinaiticus webpage[9] can access the full data from the transcription in combination with high-resolution images, users of the digital edition will be able to account for every editorial decision presented in the apparatus with visual representations of the ultimate sources. All the data are linked up and the electronic edition allows moving back and forth, from a specific reading in the apparatus to the manuscripts that testify to this reading down to the level of the relevant image of the manuscript page(s) and from a specific form of text found in one manuscript to the full spectrum of all the variant readings at that very passage in the apparatus. Hence a fully integrated digital edition!

---

[7] An electronic edition that has the mentioned features can be found under http://nttranscripts.uni-muenster.de:80/AnaServer?NTtranscripts+0+start.anv.

[8] See the contribution by D.C. Parker to the present volume.

[9] http://www.codex-sinaiticus.net/en/manuscript.aspx.

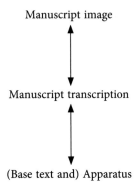

Figure 3: Integrating textual information from the many transcriptions

Conceived like that, a fully integrated digital edition is likely to change our perception and use of editions profoundly.

In the first place, our digital edition is fully accountable. The information in the apparatus can be followed back to its ultimate sources, the manuscripts themselves. The editorial work that has created the apparatus can be immediately checked for consistency and accuracy, because the full evidence is already integrated, just a couple of mouse-clicks away. Hence, editors are constantly faced with the evidence. Instead of sink or swim this means instant accountability.

Secondly, the de-contextualisation of the evidence that a textual apparatus is guilty of by its very design is made reversible. This is more than just checking whether a reading in an apparatus is erroneous or not. It is nothing less than the presence in the edition of all the original objects by means of electronic surrogates. If digitisation is remediation, it is the most transparent and context-sensitive remediation currently achievable. If we compare it with earlier remediations, like the transition from scroll to codex, this becomes immediately apparent. Texts written on scrolls are usually laid out and displayed in comparatively narrow even columns that unfold uninterrupted as the roll is opened and 'processed'. The early codices by comparison display one wide column to the page. Hence an opening offers just two wide columns, that—after turning the page—are followed by another pair of wide columns, way out of proportion to a reader accustomed to the scroll. This must have been a quite unpleasant user experience to many in the early days of introducing the codex as a vehicle for literature. It therefore comes as no surprise that the early giant parchment bibles from the fourth century experiment with a three-column (Codex Vaticanus) and even four-column (Codex Sinaiticus) layout of the codex page. Undoubtedly, the sequence of six to eight comparatively narrow columns that each of the openings consisted of[10] are also meant to recall the traditional user experience of the scroll. But these ambitious models were never adopted on a large scale. The new medium codex had a lot of advantages, but adopting page sizes that would allow for mirroring a scroll was not high on this list. By consequence, one or two-column lay-outs dominate the codex trade and a new user experience superseded the long-cherished scroll experience. The remediation of texts in the digital medium, by contrast, can in addition to the new format, contain every older format by means of a full set of images taken from the objects themselves. The different textual versions that in modern analogue editions usually exist as variant readings in a critical apparatus can now coexist with their unique material appearance, albeit as a two-dimensional surrogate image, next to the apparatus, thus allowing for a partial virtualisation of ancient reading experiences by displaying the artefacts visually on the screen. Therefore, the digital edition is indeed the most transparent and context-sensitive remediation currently achievable.

Thirdly, because the digital medium is able to convey so much more information from the witnesses themselves, electronic editions are likely to transform the way people are going to use these editions in the future. In order to take full advantage of the fully integrated digital edition, users need to develop additional skills to those demanded by traditional editions. In addition to deciphering the often idiosyncratic sigla and shortcuts used in a conventional apparatus, such new skills involve reading different types of handwriting with their abbreviations and orthographical oddities, and deciphering additional sets of apparatuses with which previous scholars and scribes have furnished the texts. The latter type of information[11] from the manuscripts is often treated inadequately in modern editions and is not readily available for comparison. On the other hand, the more images of manuscripts are available the more students can study all the aspects of the text's

---

[10] As an exception to the rule, the poetic books of the Old Testament (Psalms, Proverbs, Ecclesiastes, Song of Songs, Wisdom of Solomon, Sirach and Job) are written in a different lay-out with two columns to the page.

[11] In Biblical manuscripts we find lists of capitula (tables of content with numbered chapter headings), reading aids for reading texts in the service, and tables and numbers referencing parallel passages in the Gospels, to mention only the most prominent additional apparatuses that most manuscripts are equipped with.

appearance and familiarize themselves with the necessary skills to explore the riches of the textual tradition. Hence the digital edition will also be used as a tool to learn and teach the perception of the material aspects of human literary heritage. Indeed, through comparatively easy access to the primary sources and a broader dissemination of the skills to read and explore them users of the fully integrated digital edition are likely to become (virtual?) contributers to the edition itself. Transcriptions can be checked and improvements suggested. Users with specialised expertise can contribute additional information, such as art historians in the case of decorations and illuminations. We should therefore conceive of the fully integrated digital edition (*fide* 1.0) as an interactive platform encompassing additional modules and specific applications that allow the integration of user-generated content, making for the fully interactive digital edition (*fide* 2.0). Allowing users to contribute is not only helpful for weeding out errors. Adding new information on a larger scale—from manuscripts, versions or other sources—and new services, as well as refining analytical tools, is likely to attract specialists from other backgrounds who will consult the digital edition with different research questions and can in turn add their share of new services and additional layers of information.

Issues and obstacles: A wish list

Such a fully integrated digital edition does have to address a number of concerns, and therefore needs to meet certain standards. I have to confess from the very beginning that I may not identify all of them, let alone answer them to everyone's satisfaction.

It may be convenient to distinguish between issues that refer to the relatively new technical aspects on the one hand and issues referring to the traditional parties that are involved in producing scholarly editions on the other.

1. A fully interactive digital edition needs to have a technical infrastructure, a workspace made available so that people who are interested can actually use it. They need to know where to go and where to receive support. We are in the privileged position to have institutions that are currently developing such workspaces, for example, in the Netherlands, in Germany and in the United Kingdom. In addition it is good to know that these institutions are government-

funded bodies, subscribe to open source and open access policies and collaborate in their pioneering work. In the Netherlands there is the Huygens Instituut[12] (belonging to the KNAW) where they currently develop, *inter alia*, an on-line transcription and collation tool that supports the production of transcriptions and apparatuses in the browser. In Germany the Text Grid project,[13] funded by the Bundesministerium für Bildung und Forschung, and in the United Kingdom the Institute for Textual Scholarship and Electronic Editing[14] at the University of Birmingham are similarly developing modular platforms for collaborative editing. Solutions for similar challenges are sought from different perspectives. I am confident that the developers of such platforms will not only address issues of security and accessibility, but also, for example, the issue of accountability. Suppose some manuscript descriptions have been completed by doctoral students and form an integral part of their theses. If this material is going to be part of a larger collaborative online database, it is important to make sure that even in this large sea of data the specific subsets are still identified and kept identifiable in order to account for their role within a well-established academic process of writing a doctoral thesis.

2. Let me now turn to the traditional partners in the production chain of scholarly editions and give a list of desiderata for their contributions:

a) Holding institutions. Of course, I would like to see the major libraries digitize their manuscript holdings. I am glad to see that more and more institutions embark on such mass digitization plans. The British Library appears to be heading in that direction. Other libraries are thinking along similar lines. It is a matter of bringing the scholarly interest in such resources to their attention and talking with them about the benefits they may have from digitalizing their manuscripts, such as conservation, easy access without the actual artifact being handled, and publicity, since all the images can be watermarked to contain a reference to the holding institution. Moreover, libraries should be made

---

[12] http://www.huygensinstituut.knaw.nl/.
[13] http://www.textgrid.de/.
[14] http://www.itsee.bham.ac.uk/.

to feel proud of having their precious artifacts presented within the larger framework of scholarly digital editions.

b) Researchers. The main agents of making scholarly editions and resources pertaining to such editions are researchers, of course. Therefore, I would like to stimulate researchers who are planning editions, manuscript catalogues, checklists and similar such projects that generate a lot of data to think about the larger community they are serving with their work. The intention is not to distract them from producing splendid printed editions or catalogues. I rather want them to give up thinking that the printed book is the only medium to conveniently present the assembled information. Large sets of data simply call for digital storage and formats that make them easily convertible into databases, hence eventually usable in a digital workspace and for a fully interactive digital edition. Researchers who do not feel comfortable with deciding on data formats and structures should be encouraged to seek advice from institutions such as the Huygens Instituut, in order to ensure that their material (their data files) are properly stored and can be used by others.

c) Funding bodies. In my view, funding bodies should not only encourage researchers to care for the digital sustainability and usability of the data that are generated in the course of their projects. Funding bodies should make it a formal requirement in application procedures for such projects that the applicant explains how he or she will ensure the digital sustainability and usability of their data. I think that the proposed output in such familiar forms as 'printed edition', 'printed catalogue', etc. is not enough. To be very explicit, and likely controversial, I encourage funding bodies to turn down applications for making editions, catalogues and similar products that are silent about the digital set-up and storage of their data. In return for spending public money, the public is entitled to expect good value for money. And the bottom line is that the data generated in the course of such projects are digitally preserved and made available for other projects.

d) Publishers. Publishers traditionally perform the service to make scholarly products available. This service is not free, but has to be paid for. In order to generate money the publishers take over the copyright from the author and make that the foundation of their business model. If somebody wants to access scholarly information published in a book or scientific journal, he or she has to pay the publisher for it. When it comes to huge and complex products like critical editions of the Bible, the scholarship that is finally displayed in such an edition consists of the work of generations of paleographers, cataloguers, classical and Biblical scholars of various expertise. They all have assembled, checked and filtered a massive iceberg of data of which the final edition presents to the user the mere tip. And as a result of that, much information has already been published and continues to be published in one or the other way, for example in manuscript catalogues, checklists, editions of individual manuscripts or groups of manuscripts. Not only researchers and funding bodies should contribute their share of awareness towards digitally maintaining the data generated through the mentioned types of research. But also publishers should be made aware of the potential drawbacks rigidly applied copyright restrictions can have on future research in that area. Publishers who sell any of this partial information—partial from the perspective of the big critical edition—should adopt a stance that leaves the copyright of the data *behind* the products they print with the public that usually funds such work.[15] Collaboration, building upon the work of others and integrating data for more comprehensive analysis should not be hindered because fractions of the material have been printed by different publishers. It should be obvious to the publishers that the more material is digitally accessible and seamlessly integrated, the more research questions can be addressed, which results not only in more printable analysis of the material, but will attract more people to the field. Ideally, a fully interactive digital edition is of service to all partners.

References

Gamble, Harry Y., *Books and readers in the Early Church. A history of Early Christian texts* (New Haven: Yale University Press, 1995).
Hamel, Christopher de, *The Book. A history of the Bible* (London: Phaidon Press, 2001).
West, Martin L., *Textual criticism and editorial technique applicable to Greek and Latin texts* (Teubner: Stuttgart, 1973).

---

[15] Cf. the Creative Commons initiative at http://creativecommons.org/index.php.

DISTRIBUTED NETWORKS WITH/IN TEXT EDITING AND ANNOTATION

Vika Zafrin

Though the persistent image of a humanities scholar is that of someone working in isolation, the social aspects of her work should not be underestimated. Certain mediums for online interaction have been gaining popularity in the humanities. Blogs are one such medium; many not only allow but encourage other to make detailed commentary on the authors' thoughts. Similarly, online articles that allow commentary have been popular with both academics and the general public. Software is increasingly available that enables people working in geographically disparate areas to work on documents collaboratively in real time over the internet.

In this essay I examine several commenting engines that have been developed and/or used by humanities scholars for text editing and annotation. I discuss factors that influence decision making in the creation of such tools; how they have been used by humanists; critical feedback produced by those scholars; and relevant recently funded projects. The primary foci for the essay will be Brown University's Virtual Humanities Lab; the CommentPress engine; a commenting engine used by the Free Software Foundation; and the scholarly research tool Zotero, built at George Mason University in the US. I then address some of the concerns raised thus far regarding the online publication and archiving of this form of knowledge work, and argue for its crucial role in today's humanities research.

Connecting (and re-connecting) with colleagues at conferences and colloquia such as the one whose papers are collected in the present volume is possibly my favorite aspect of attending these events. At their best, they can be both relaxing and electrifying, a recharge, an inspiration. We give and hear papers, and often spend time after hours engaging in further dialogue. We take advantage of the time available, because no method of communication provides the constant feedback of face-to-face. We network.

Then we get home, and here is where practices among humanists have been diverging in recent years. The conventional image is:
a humanities scholar arrives home (or is already there, because she
doesn't 'do' conferences in the first place); she sequesters herself in an
office that functionally resembles a monastic cell; she reads and writes
alone, emerging to teach, visit the library, and get coffee. Occasionally
she will publish articles and books, and in this way communicate with
scholarly communities relevant to her research. That is, she will succeed in communicating with relevant colleagues if they happen to find
her publications and read them.

That is a common view of how humanists get their work done. Of
course, the reality is often different; in the past decade or two this difference has been increasingly stark. Contemplative solitude still has its
place; but available means of electronic communication have become
so numerous, and so ubiquitous, that if you are on the internet at all, it
is difficult not to encounter numerous opportunities for collaborative
work and other forms of online scholarly communication.

The number, variety and nature of venues for this is not necessarily an advantage. It can be difficult to find enough time for in-depth
email correspondence. Scholarly weblogs are on the rise, but unless
one is extremely selective with regard to how many and which of them
to follow, they can quickly fill up more hours than we tend to have
available. The amount of new information published on the internet
daily is overwhelming.

Yet networked communication presents us with opportunities we
simply did not have before. It is becoming significantly easier to find
fellow scholars interested in the same obscure topics that we pursue,
even if those scholars are working halfway around the world. It has
often been said, particularly in discussions pertaining to institutional
repositories, that openly publishing one's research on the internet—
whether formally or informally—gets one cited more often.[1]

But it is not enough to put new content online. The content must be
discoverable by the right people in an ocean of information, and this
is where distributed, collaborative annotation comes in. In this essay
I will set aside synchronous tools available for such activities—tools

---

[1] For a list of relevant studies, see the Open Citation Project. The effect of open
access and downloads ('hits') on citation impact: A bibliography of studies. http://
opcit.eprints.org/oacitation-biblio.html (accessed 29 January 2009).

that require people to be online at the same time: these are very useful, but it is often difficult to synchronize schedules on a consistent basis. I will address instead networked annotation of texts as asynchronous scholarly activity.

Some definitions

For the purposes of this essay, I will define both *text* and *annotation*, both of which I am interpreting rather widely and in the specific context of the network. A *text*, then, is any artifact with a URL or other unique address, that can be linked to and regarded as an object of critical analysis. Originally paper-bound, digitized primary objects of study such as philosophical treatises, legislative documents and poetry are one type of text. Another broad category includes non-verbal art—digitized paintings, music, maps. And a third important category includes born-digital primary objects such as topical blog posts, software and games.

Annotation is a critical addition to a text, usually attached to a specific part of that text, which provides *notes*—additional information, usually falsifiable, that fleshes out a fact or reference. For the purposes of this essay, annotations are more formalized and strictly informative than *comments*, which in turn are more discursive and less systematic. However, the differences between the two words are subtle, and they are often used interchangeably except in some specific fields (notably medieval literary studies). For the purposes of this essay, comments are a sub-category of annotations, less formal and more casual than other annotation types.

As with text, annotation can take a variety of forms. One of them is what humanists regard as traditional scholarly annotation—commentary on a text or part of one, in digital form. A fantastic example of this is the Princeton Dante Project (PDP).[2] Begun in 1997, the PDP presents all of Dante's oeuvre online. The *Divine Comedy* in particular is presented with line-by-line annotations indicated by special symbols placed beside each verse, and accessible by clicking on those symbols. The annotations themselves are original, but synthesize and address past interpretations by well-known scholars and writers.

---

[2] http://etcweb.princeton.edu/dante/ (accessed 29 January 2009).

The project does not stop there; influential past commentaries are accessible through hyperlinks within these first-level annotations. Clicking on a reference link will take the reader to the Dartmouth Dante Project,[3] a sister project of the PDP, which 'combines modern information technology with nearly seven hundred years of commentary tradition on Dante's major poem, the *Commedia*,' and aims to 'put the entire texts of more than 70 commentaries into a searchable database that anyone can access via the World Wide Web.'[4] The two projects are seamlessly integrated, and continue to present the most comprehensive extant collection of Dante's texts and commentaries on them.

Social tagging is another form of annotation. Using services such as Delicious,[5] which claims to be 'the biggest collection of bookmarks in the universe,' users not only bookmark digital objects and share them via syndication feeds, but also assign keywords (tags) to them, effectively indexing the internet. Such large-scale efforts are often rightly regarded with suspicion by academe, but as with any authored information source, one can easily choose to follow only trustworthy authors, groups and other entities, whose reputation is built gradually, based on actual contributions to the knowledge base.

Still other forms of annotation are associated with weblogs—publishing platforms for articles and journal-style entries, usually presenting an opportunity for readers to comment on them. Most comments on blog posts fall into two broad categories. On one hand are comments that engage blog post authors and readers in conversation—or try to provoke a reaction by being inflammatory, a common feature of conference question-and-answer sessions. On the other hand are comments that provide additional information or otherwise do not call for a response.

In addition, there are tools—such as CommentPress, which I will discuss in depth below—designed specifically for authors who wish to publish materials on the net and solicit paragraph-by-paragraph commentary on them. Such commentary, while not functionally different from any other format in which peer review takes place, nevertheless offers peer review in a networked social space, where such tools have

---

[3] http://dante.dartmouth.edu/ (accessed 29 January 2009).
[4] http://dante.dartmouth.edu/about.php (accessed 29 January 2009).
[5] http://delicious.com/ (accessed 29 January 2009).

been unavailable until recently. This may not change the nature of the work performed, but it changes the nature of *how* humanities scholars work, and as such is worthy of consideration here. Let's look at some concrete examples of networked text editing and annotation.

Virtual Humanities Lab

From 2004 to 2006 I was Project Director of the Virtual Humanities Lab (VHL),[6] a two-year experimental project co-funded by Brown University and the U.S. National Endowment for the Humanities. VHL was housed at Brown, but had collaborators throughout North America and Europe. Principal Investigator Massimo Riva and I, along with our colleagues, envisioned it as an online laboratory for collaboratively editing and annotating texts. We conceived the annotation tool as being useful beyond the scope of the project, and the site as a nexus of text analysis in Italian studies and related fields.

Our ultimate output was twofold. First, we worked with philologists and historians on the semantic encoding[7] of three Italian texts from the early Renaissance, seminal to the nascent Humanist movement: Giovanni Pico della Mirandola's famous *Conclusiones Nongentae Disputandae* (the *900 Theses*); Giovanni Boccaccio's *Esposizioni* (*Expositions on Dante's Divine Comedy*); and Giovanni Villani's *Cronica Fiorentina* (the *Florentine Chronicle*), a history of Florence up until 1348, when the author died of the Plague. In parallel with this encoding activity, we created an annotation engine for these texts, as well as for Boccaccio's *Decameron*, an encoded version of which we already hosted on the *Decameron Web*.[8]

VHL was conceived as what Paul Eggert in 2005 described as a 'work-site,' a web-based space for collaborative editing and annotating of texts.[9] 'We can encode only what we know about texts,' Eggert notes. 'Over time, there will be expansion of what we know, correction

---

[6] http://golf.services.brown.edu/projects/VHL/ (accessed 29 January 2009).
[7] Semantic encoding is the practice of inserting metadata into a document, usually using special tags enclosed in <angle brackets>. Semantic encoding differs from structural encoding (<p> for paragraph) in that any concept can be encoded. These examples are <word type="adjective">trivial</word>, but <idiom lang="english">get the point across</idiom>.
[8] http://www.brown.edu/decameron (accessed 29 January 2009).
[9] P. Eggert, 'Text-encoding, theories of the text, and the "work-site"'.

and reformulation. [...] [C]ollaborative interpretation, accumulating over time, contributed by scholars and other users must be consciously allowed for and enabled.' (430) In the process of encoding, we acutely felt Eggert's point about an evolving understanding of texts.

As our encoding language we used XML, the eXtensible Markup Language. Aside from being the markup language of choice for many computing humanists, XML's main advantage is its structural simplicity. There are only a few rules, which, when adhered to consistently, make an XML document *well-formed*:

- Most encoding consists of elements which may or may not contain attributes: <fruit type="apple">Granny Smith</fruit>;
- Capitalization matters: <this> is different from <This>;
- Elements must be properly closed: <example>like this</example>; and
- Elements must be properly nested: like <outer> dolls in which <inner>another doll </inner> may be nested</outer>.

We asked all three of our encoders to perform their work without a DTD (Document Type Definition, a template that defines the tagging structure of a document written using a markup language such as XML), using whatever elements they deemed appropriate, as long as the XML was well-formed. There were only two specific guidelines to follow:

- that the categories elucidated by the encoding be broad enough to produce interesting search results, and
- that, in their estimation, colleagues interested in these texts would generally find their encoded aspects interesting as well.

We decided to postpone the creation of a DTD primarily because none of the three principal encoders had ever done this sort of work, not to mention encoding on such a large scale. The encoding scholars received training in basic XML and in the use of oXygen[10] software for XML encoding, as well as the Subversion revision control system[11] for

---

[10] http://www.oxygenxml.com/ (accessed 29 January 2009).

[11] http://subversion.tigris.org/ (accessed 29 January 2009). A revision (or version) control system allows management of electronic file versions as the files are edited. Common features of such systems are the ability to revert to an earlier revision,

sharing work remotely. (All three encoders were working at different institutions at the time.) Thereafter we left them to work on their own, making help available upon request.

Most importantly in this experiment, we wanted our encoders to run with the concept of semantic encoding—a form of annotation—at its most basic, and see what happened. The process ended up being highly iterative. There was a lot of back and forth communication. In the beginning discussion concerned mostly what was to be annotated in the first place; later it moved on to the structure of the annotations— what was to be an element, what attributes were the most important.

We used networked communication for text editing and annotation from the very beginning (see remote access above). Our two principal encoding projects, the *Esposizioni* and Book XIII of the *Cronica*, were around 700 and 200 modern print pages in size respectively. Two collaborators encoded the *Cronica*; the *Esposizioni* had one principal encoder and several researchers investigating specific issues.

After first-pass encoding was completed, we performed a substantial amount of cleanup work, usual for such projects—tightening the encoding structure, catching inconsistencies and typos inside tags, and making sure that the result was well-formed. Most of this work was performed by myself; this was more efficient than it would have been for the encoders to learn about regular expressions[12] and other tools that made the job substantially easier.

This phase was the equivalent of copyediting, with one important difference: because everyone had remote access to the texts, communication regarding it happened directly inside the encoded files. I was able to correct obvious typos myself, but there were many instances in which I had to ask the scholars for clarification. Instead of calling or emailing them with questions, I inserted comments[13] in appropriate places. Later the encoders were able to open the files, find those

---

branching (making two child revisions of a parent file, and maintaining their development in parallel) and integration of changes made by two or more authors around the same time.

[12] Regular expressions (regex) make searching large texts easier. Using special syntax, one may search, for example, for accidentally inserted spaces between words and punctuation, and eliminate them. Using regex to copy-edit makes the process go much faster than it otherwise would, and may reduce the human-error factor.

[13] In a computer-readable file, a comment is usually text enclosed in special tags that make the computer ignore it, and neither print it nor consider it part of the code.

comments, and make corrections directly in the files—and so we eliminated the additional step of sending modifications back to a central editing point. Since we continued this work after publishing the texts on the web, the results of such modifications were immediately visible on the website.

The result for each of these texts was, functionally if not expressly, the beginning of a scholarly edition. The collaborative process of creating this encoding—the encoding we ultimately employed to build a number of indexes[14]—was possible because we used the network for versioning and communication.

In parallel with the above-described work, we collaborated with Brown's Scholarly Technology Group[15] to build and later utilize an annotation engine. This was before CommentPress[16] was released; our purpose was to provide a tool that scholars could use to annotate specific parts of the texts we had placed online, and to discuss controversial passages. After a prototype and some beta testing, we came up with sometjing that looks like the screenshot shown in Figure 1.

Electing to view part of a text hosted on the site yields a view of that section with highlights that reveal some of the semantic encoding (which will have become part of the new digital object, if not of the base text), and indicate whether any other annotations (external to the original base text) are already associated with it. If there are, users can elect to view them or not, and registered users may make new annotations, as well as edit or delete their own annotations. The engine allows for an anchor word or phrase from the main text to be associated with the annotation; annotations themselves may contain basic HTML; and, importantly, each annotation may be attached to more than one text segment, for intra-textual comparison and other cross-referencing work.

The engine is far from perfect, and has already been eclipsed in many respects by tools built concurrently with, or after, our work on it was done. When we started, however, there was no openly available tool that did what we wanted to do.

---

[14] For the *Esposizioni*, we created indexes of people, places, themes, mythological characters and literary works mentioned in the text. The *Cronica* currently has indexes of people and places. Clicking on any item in an index brings up a contextualized listing of all its occurrences in the text.

[15] http://www.stg.brown.edu/ (accessed 29 January 2009).

[16] http://www.futureofthebook.org/commentpress/ (accessed 29 January 2009).

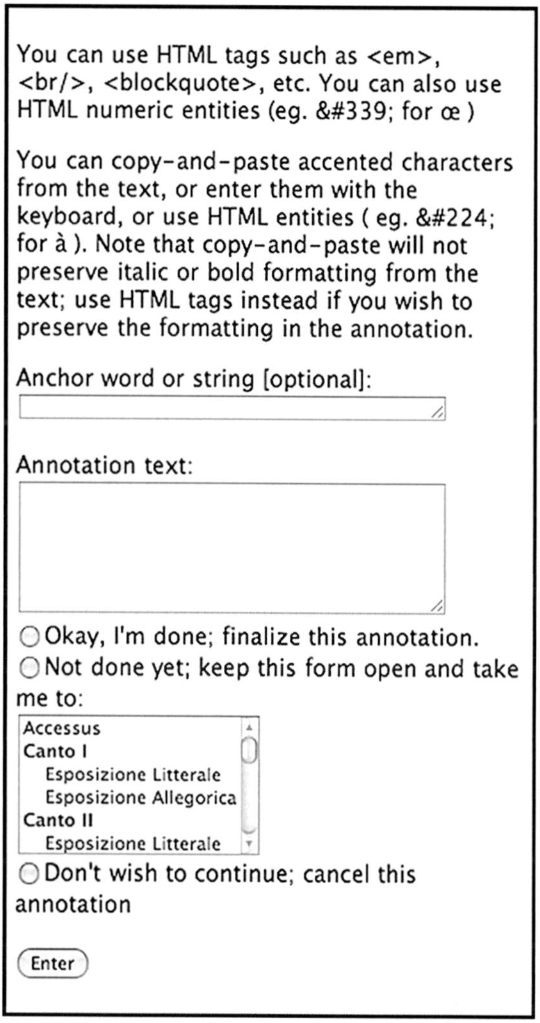

You can use HTML tags such as <em>, <br/>, <blockquote>, etc. You can also use HTML numeric entities (eg. œ for œ)

You can copy-and-paste accented characters from the text, or enter them with the keyboard, or use HTML entities (eg. à for à). Note that copy-and-paste will not preserve italic or bold formatting from the text; use HTML tags instead if you wish to preserve the formatting in the annotation.

Anchor word or string [optional]:

Annotation text:

○ Okay, I'm done; finalize this annotation.
○ Not done yet; keep this form open and take me to:

> Accessus
> Canto I
>   Esposizione Litterale
>   Esposizione Allegorica
> Canto II
>   Esposizione Litterale

○ Don't wish to continue; cancel this annotation

(Enter)

Figure 1: Annotation engine developed in collaboration with Brown's Scholarly Technology Group

Several dozen contributors, among them several internationally known senior scholars, indicated interest in collaborative annotation of the texts we had put up. Of course, garnering interest and getting people to actually contribute are different tasks. While we did receive some contributions, most scholars—particularly those who do not spend significant time on the internet as part of their workflow—just did not have the time or energy to repeatedly return, look for new contributions, and add more annotations.

We had known that, given the available time and resources, we would need to apply for another grant in order to render the site more likely to be used for its intended purpose. Even before the two-year grant period was concluded, we already had a wish list of features to add: the ability to see all of a particular author's contributions; the ability to see the most recent contributions; discussion threading, so that annotations could serve as replies to specific other annotations, and be hierarchically presented much like newsgroup or email threads; the ability to view and discuss the semantic encoding—which, being a form of scholarly annotation, needs peer review; and finally, the ability to use the annotation engine with other texts, perhaps encoded by other people and housed on other servers. In short, in order to be truly useful, the annotation engine needs to be a stand-alone tool that is relatively easy to install and run.

Most importantly, however, in order for such a project to succeed, a crucial component is a dedicated managing editor and outreach coordinator. This would be a person who gathers interested scholars, rallies the annotating troops, and continually spurs conversation. Without such a constant external stimulus, the dozens of researchers scattered around the world will simply not pursue the project, and the network's collaborative opportunities will continue to be glacially slow in their incorporation into humanities scholars' workflow.

Grand Text Auto, Expressive Processing *and CommentPress*

Let us look at another example of collaborative text annotation. CommentPress is an optional component for the popular WordPress[17] blogging engine. It was developed by the Institute for the Future of the Book (if:book),[18] and provides the ability for readers to comment

---

[17] http://www.wordpress.org/ (accessed 29 January 2009).
[18] http://www.futureofthebook.org/ (accessed 29 January 2009).

on a blog post paragraph by paragraph, engaging in more detailed discussion than post-level commenting easily allows. Comments are also threaded, and conversations thus easier to follow (see Figure 2).

In the course of his work on *Expressive Processing*, forthcoming from MIT Press, Noah Wardrip-Fruin successfully used[19] this commenting engine to solicit comments on the group blog in which he participates, Grand Text Auto (GTxA).[20] Before Wardrip-Fruin's experiment, CommentPress had been a 'theme' for WordPress, meaning that it could be applied to the entire website, or not at all. In January 2008, if:book[21] wrote up the story of how, through collaboration between WordPress

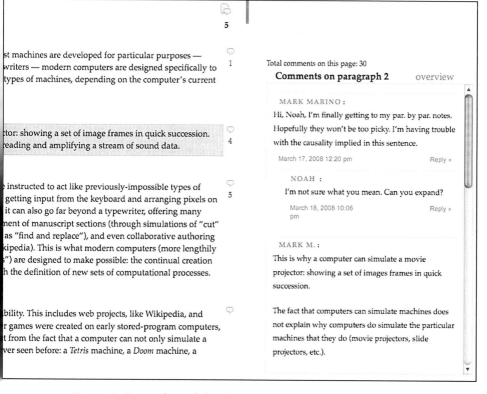

Figure 2: Screenshot of the Comment Press annotation plugin

---

[19] http://grandtextauto.org/category/expressive-processing/ (accessed 29 January 2009).
[20] http://grandtextauto.org/ (accessed 29 January 2009).
[21] See http://www.futureofthebook.org/.

developer Mark Edwards and scholar-programmer Jeremy Douglass, CommentPress became a plugin.[22] GTxA was able to use it with posts related to *Expressive Processing*, but turn it off on the rest of the site.

This had a subtle but important effect. GTxA is not usually a peer-review-oriented site; the discussion on it is often more generalized than would be appropriate for the CommentPress engine in terms of eventual readability of comments. A certain format has by now evolved for blog posts and subsequent comments on them, and audiences expect that format unless there is a good reason for it to change. So it is understandable that the GTxA authors did not convert the entire blog to CommentPress. Yet, the community already established there by early 2008 was an essential part of Wardrip-Fruin's experiment. He could have set up an entirely different website for peer review, but would have probably lost community members (and their input) in the transfer process. By modifying CommentPress to work seamlessly with the established blog no additional barrier to participation was set up, which was a significant boon to motivating contributions.

Altogether, there are 458 comments related to *Expressive Processing* on Grand Text Auto. Many of them are Wardrip-Fruin's own, as he actively participated in discussion with his readers; but the sheer amount of discussion—which took place over a period of roughly six months—is still impressive. In April of 2008, Wardrip-Fruin wrote up his impressions of the CommentPress peer review experience. Here is what he said:

> In most cases, when I get back the traditional, blind peer review comments on my papers and book proposals and conference submissions, I don't know who to believe. Most issues are only raised by one reviewer. I find myself wondering, 'Is this a general issue that I need to fix, or just something that rubbed one particular person the wrong way?' I try to look back at the piece with fresh eyes, using myself as a check on the review, or sometimes seek the advice of someone else involved in the process (for example, the papers chair of the conference).

> But with this blog-based review it's been a quite different experience. [...] Personally, I didn't foresee it. I expected to compare the recommendation of commenters on the blog and the anonymous, press-solicited

---

[22] http://www.futureofthebook.org/blog/archives/2008/01/expressive_processing_an_exper.html (accessed 29 January 2009).

reviewers—treating the two basically the same way. But it turns out that the blog commentaries will have been through a social process that, in some ways, will probably make me trust them more.[23]

I will come back to this at the conclusion of this essay. In the meantime we move on to consider a completely different type of commenting tool.

GNU Public License Commenting Engine

When the Free Software Foundation[24] was developing a new version of the GNU General Public License[25] in 2007, it opened the text of the new license for comments. The comments were gathered using a simple and elegant engine built for the FSF by the Software Freedom Law Center.[26] This version of the license has now been published, and comments are no longer being accepted, but it is still possible to view the text and existing comments via an impressively subtle user interface.[27]

The web page referenced in that last footnote mostly contains just text, with no fancy graphics or layout features. The entirety of instructions to commenting readers is: 'select some text and then type "c" to submit comments.' The reader's input is automatically linked to the selected section. Commented text segments have a highlighted background, which changes color from pale yellow through red to black depending on how many comments it received. Mousing over a highlighted section reveals how many comments are attached to it; clicking on it opens those comments to the right of the text, obscuring no part of it.

The GNU license is a widely used one, so comments not only were numerous (298) but often quite long. The comment display engine automatically cuts off each long statement and makes it expandable and collapsible with a single, short, obvious link. In addition, in order to cut down on redundancy, readers who agreed with a particular

---

[23] http://grandtextauto.org/2008/04/05/blog-based-peer-review-some-preliminary-conclusions-part-2/ (accessed 29 January 2009).

[24] http://www.fsf.org/ (accessed 29 January 2009).

[25] http://www.gnu.org/licenses/gpl.html (accessed 29 January 2009).

[26] http://www.softwarefreedom.org/ (accessed 29 January 2009).

[27] http://gplv3.fsf.org/comments/gplv3-draft-4.html (accessed 29 January 2009).

comment were able to register that without writing something similar to what had already been said. In that case at the end of comments a short note appears in square brackets, stating how many agreements have been registered.

It is surprising that this commenting engine is not used more widely. It allows for lengthy discussions yet isn't cluttered; it provides immediate visual cues as to which sections of the document are hotly debated (although I will note that it makes no provision for color blindness); and it is as inviting as possible to newcomers: the commenting instructions are one line long. Distributed networking, when the nodes are scholars already busy as a result of doing a lot of research and thinking, seems to be as functional as the barriers to entry are low; this is even more the case for online communities, where there is an expectation that people will return repeatedly, perhaps even on a regular basis. Given that, it will be exciting to see whether this or a similar system takes off in academe.

Social tagging as text annotation

As I mentioned before, texts can be anything addressable, and some are born-digital objects. It follows that the product of social tagging sites qualifies as annotation. I have already touched on Delicious; another interesting example to look at is Diigo.[28] Short for *Digest of Internet Information, Groups and Other stuff*, Diigo is a personal research tool that works as a plugin for the Firefox, Internet Explorer and Flock web browsers. Upon registering for a free account, users may highlight and privately annotate portions of web pages. Comments that apply to entire pages, as well as tags, are also available features, and can be made public. When the Diigo plug-in is installed and active on one's browser, viewing a page with any public annotations made by Diigo members will reveal those annotations automatically, in the Diigo Sidebar (optionally made visible or hidden) or directly on the page. There are groups to join or follow, according to user preferences—Diigo explicitly describes itself as a 'social information network (SIN).'[29]

In other words, Diigo is a service similar to Delicious, but with one important difference: the emphasis is on annotation, not bookmark-

---

[28] http://www.diigo.com/ (accessed 29 January 2009).
[29] http://www.diigo.com/about (accessed 29 January 2009).

ing. The result is less a referral or discovery service than value addition to web pages that a user is going to view regardless. Discovery, however, is also possible through the social networking tools common to such sites ('friends' lists that allow one to follow particular individuals' input; and available lists of users new to Diigo, or featured users).

Zotero

It would be difficult to survey all available digital tools for collaboration in the humanities; but some do stand out. Among them is an eagerly anticipated coming attraction—Zotero 1.5, currently out in beta. Zotero[30] is a scholarly research and annotation tool developed at the Center for History and New Media at George Mason University in the U.S. It burst onto the scene in 2006, and is now used by over a million individuals.[31]

Like Diigo, Zotero is a browser plugin. It provides a library in which to collect research-related objects from web bookmarks to text and multimedia files to snapshots of entire web pages. Objects can be grouped in a variety of ways, playlist-like, and declared to be related to each other; notes and tags can be attached. There are built-in citation tools and many supported export formats, as well as interoperability with popular office suites and the WordPress blogging platform. The interface is available in over 30 languages. There even exist plugins for the main Zotero plugin; a notable one is Vertov,[32] which allows annotation of media files. The entire project is free and open source, and is backed by major funding.

Zotero 1.5 Sync[33] allows scholars to store their data on a centralized server, rendering it accessible from multiple locations. It is also possible to make one's Zotero library public; crucially for our purposes, notes made on library objects may also be publicly shared (this is an option separate from sharing a library; in other words, a scholar's notes may remain private even as she makes her library public). Users

---

[30] http://www.zotero.org/ (accessed 29 January 2009).

[31] Dan Cohen, co-Director, Zotero Project. Personal correspondence, 29 January 2009.

[32] http://digitalhistory.concordia.ca/vertov/ (accessed 29 January 2009).

[33] http://www.zotero.org/support/sync_preview (accessed 29 January 2009).

have requested the ability to share only certain collections within their libraries, and this functionality is currently in development.[34]

This presents collaborative opportunities for analyzing anything with a web address. A researcher may take notes in the course of her own work, intended for herself: this process is thus already in her usual workflow and, aside from being performed within Zotero and shared, does not require any additional effort. Another researcher may happen to read her publicly available notes, and perhaps share his thoughts via notes in his own collection, or by email or other communication means.

Zotero may certainly be used for intentional and lengthy conversations about texts. However, its rare appeal in the above hypothetical scenario is that, while a collaborative endeavor may well grow out of quotidian work, it is not set up as such in advance. Scholars do not commit to what they would view as an addition to their work load, which (like CommentPress in GTxA) pre-empts a potential obstacle to participation. Such collaboration is spontaneous and unstructured; Zotero is one of the few venues that allow for this kind of work.

Recent developments

Collaboration has been a popular buzzword in many areas of the humanities, including and beyond text editing and annotation. As digital humanities projects become increasingly visible, and the technological opportunities more interesting, we call on experts in many other fields to help us make our projects a reality. The trend of collaboration among humanists, librarians and technologists[35] has been encouraged by recent funding opportunities, and reflected at digital humanities conferences.[36]

---

[34] http://forums.zotero.org/discussion/5861/can-i-publish-selected-collections-instead-of-publish-entire-library/#Item_0 (accessed 12 March 2009).

[35] In reality, professionals working in digital humanities often do not fit into discrete groups, but are experts along a spectrum spanning these and other areas of inquiry.

[36] Working as I do in the United States, I am most familiar with funding opportunities and projects in that country. Work relevant to collaborative text editing and annotation is, however, being pursued worldwide. For an extensive annotated list of digital text analysis tools available to humanities scholars, please see http://digitalresearchtools.pbworks.com/Text-Analysis-Tools (accessed 6 August 2009).

The U.S. National Endowment for the Humanities has been making great strides toward encouraging digital innovation in the humanities. Following are a few of the recently funded projects that will be of particular interest in the current context.

- The Carolingian Canon Law Project at the University of Kentucky seeks to establish 'encoding standards and digital access for multiple versions of medieval Latin legal manuscripts, including bibliographic information, annotations, and English translations.'
- Haverford College's 'The Chansonniers of Nicholas Du Chemin (1549–1551): A Digital Forum for Renaissance Music Books' will offer 'a searchable image archive, an innovative electronic display for music books, commentaries and examples, and tools for research, transcription and collaboration.'
- Independent scholar Michael Newton's 'Building Information Visualization into Next-Generation Digital Humanities Collaboratories', his abstract says, 'will allow multiple users to contribute to, discuss, edit, and utilize a common body of information.'
- 'Digital Tools/AXE' at the University of Maryland, College Park, will 'facilitate collaboration in the digital humanities by permitting multiple scholars to work on the same document or archive at the same time from various locations.'[37]

Larger-scale endeavors are also gathering momentum. Project Bamboo, jointly based at the Unviersities of California and Chicago, is 'a multi-institutional, interdisciplinary, and inter-organizational effort that brings together researchers in arts and humanities, computer scientists, information scientists, librarians, and campus information technologists to tackle the question: How can we advance arts and humanities research through the development of shared technology services?'[38] On an international level, centerNet—'an international network of digital humanities centers'—is consolidating and promoting efforts in the creation of a global cyberinfrastructure, which will in turn promote connections and collaboration.

---

[37] More information on these and other projects funded by the NEH Office of Digital Humanities can be found in the Library of Funded Projects at http://www.neh.gov/odh/ (accessed 6 August 2009).
[38] http://www.projectbamboo.org (accessed 6 August 2009).

The international Digital Humanities conference, put on by the Alliance for Digital Humanities Organizations (ADHO),[39] has seen an increasing number of presentations relating to collaborative endeavors in text-based humanities research. Consider the following paper titles from the 2009 conference, hosted by the University of Maryland:[40]

- 'No Job for Techies: Collaborative Modelling as an intellectual activity of the analyst and scholar in the development of formal representations of scholarly materials' (John Bradley, King's College London);
- 'Africa Map Release I Beta: An Infrastructure for Collaboration' (Suzanne Blier, Peter Bol, Benjamin G. Lewis, Harvard University);
- 'Library Collaboration with Large Digital Humanities Projects' (William A Kretzschmar, Jr., William G. Potter, University of Georgia);
- 'Supporting the Creation of Academic Bibliographies by Communities through Social Collaboration' (Hamed Alhoori, Omar Álvarez, Miguel Muñiz, Richard Furuta, Eduardo Urbina, Texas A&M University).

Clearly, there are trends here. One of them is large-scale and organizational; the other is much more topically specific. Both kinds of efforts are very practical. Collaborations in the digital humanities are developing so rapidly that, by the time this book is published, dozens of newer projects large and small will have begun, or been publicized.

Some tentative conclusions

It seems that scholars are incorporating distributed networking into their work in ever more creative ways. The ability to deliver information to any interested party, and also to have it delivered to us from a customizable list of contacts, addresses a lot of the hesitation researchers have about the credibility of information on the internet.

So with all this reading and writing—and annotating—as forms of digital scholarship, what does the popularity of these tools and

---

[39] http://www.digitalhumanities.org/ (accessed 6 August 2009).
[40] For more information, including a detailed conference schedule, please see http://www.mith2.umd.edu/dh09/ (accessed 6 August 2009).

approaches tell us about how research is changing? I think the news is encouraging on several complementary fronts.

Most obviously, perhaps, the multitude of venues for self-expression and interaction that these technologies create make it more likely that geographically scattered researchers whose specific interests coincide will find each other. The mode of interaction that is most encouraged is conversation; and so serendipitous connections of the non-obvious kind are more likely. In addition, academic conversation conducted and witnessed on the internet is more loosely tied to particular disciplines than face-to-face conferences tend to be. So intellectual cross-pollination—not least, as also discussed elsewhere in this volume, that among humanists and computer scientists—is facilitated by the tools and methodologies explored in this essay.

This brings us to the issue of decreased disciplinarity, perhaps the most important recent change in humanistic research encouraged by distributed networking. In 1690 John Locke wrote about the home-steading paradigm of land distribution:

> Though the earth, and all inferior creatures, be common to all men, yet every man has a property in his own person: this no body has any right to but himself. The labour of his body, and the work of his hands, we may say, are properly his. Whatsoever then he removes out of the state that nature hath provided, and left it in, he hath mixed his labour with, and joined to it something that is his own, and thereby makes it his property.[41]

The open-source software development community has embraced this paradigm, and created a reputation-based economy of knowledge sharing. Reputation is built up through actual contributions to software, and is based on whether these contributions prove useful to the users of that software. The usefulness factor is crucial to quality control, a matter of concern to online scholarship. Value is determined pragmatically, which renders lack of credentials irrelevant and barriers to entry low. Credentials themselves turn out to be both beneficial and impermanent: beneficial in that in a reputation-based economy, pre-existing reputation-based credentials are as easily accepted as academic titles by the appropriate communities; and impermanent in that, if a contributor stops participating in a community *and* her

---

[41] John Locke, *Second treatise of government*, Ch. 5, sect. 27.

previous contributions are no longer used, her credentials begin to
lose their meaning. Nobody coasts on past laurels.

All of this translates rather neatly from open-source software development to the digital humanities. Horizons of expertise are broad-
ened by necessity. University culture is changing to echo that of the
medieval European universities, in that the hierarchical distinctions
between teachers and students are softened. And although conferred
degrees do not expire, they may also lend themselves less to long-term
complacency now than they may have in the past. Venues for online
participation in scholarly communities are quickly increasing in both
number and popularity, and in the process changing the importance
of continued community involvement to a researcher's career.

Individual considerations aside, distributed networking tools are
also the best collaboration aids available to us for fully engaged discussion of born-digital objects. These tools were built, and exist, in
the same medium in which those objects were originally created; no
translation is involved at any point in the creative or critical processes.
The last decade has seen an explosion of cultural artifacts that are born
digital, and their number will continue to increase exponentially. If
humanities scholars wish to continue to study all of the humanities in
their varied forms, we might as well stop 'dancing about architecture'
and treat digital artifacts in their native habitat, using all the richness
of the digital medium.

References

Eggert, P., 'Text-encoding, theories of the text, and the "work-site"', *Literary and lin-
     guistic computing* 20, 4 (2005), 425–36.
Locke, John, *Second treatise of government* (1690) (http://oregonstate.edu/instruct/
     phl302/texts/locke/locke2/locke2nd-a.html; accessed 29 January 2009).

PART FOUR

WIDER PERSPECTIVES ON DEVELOPMENTS IN
DIGITAL TEXT SCHOLARSHIP

THE CHANGING NATURE OF TEXT:
A LINGUISTIC PERSPECTIVE

David Crystal

How does digitally mediated communication alter our notion of text? There are some continuities with traditionally spoken and written text, but also important discontinuities. Differences with speech include new patterns of turn-taking, the use of emoticons, and new conversational rhythms. Differences with writing include issues of persistence, animateness, hypertext linkage, and framing. A pragmatic perspective brings to light new kinds of text, such as those which include features to defeat spam filters or to ensure a high search-engine ranking, or those which raise ergonomic or ethical issues. Digitally mediated communication also raises the question of how to handle texts whose boundaries are continually changing, as in forums and comment postings. Issues of responsibility arise relating to who is the author of a text, especially in contexts where there is moderation or interactivity (as in wikis).

Introduction

To the linguist, the world of digital communication presents an intriguing and challenging research domain. It hasn't even got an agreed name yet. A bewildering array of popular and scholarly labels have been used, such as cyberspeak, Netspeak, and electronic discourse. *Computer-mediated communication* seemed to be becoming a standard, fostered by such journals as the *Journal of computer-mediated communication*, but patently this is now too narrow, in view of the emergence of devices which present text but which are not computer-mediated in the usual sense, such as mobile phones, blackberries, satnavs, and voice-interactive washing-machines. *Electronically mediated communication* is now receiving some use, most recently in Baron (2008). *Digitally mediated communication* also suggests itself; and in view of the emphasis on 'digital text' in this book, I shall stay with that.

Linguistic challenges

There are several properties of digitally mediated communication which constitute a challenge to linguists wanting to explore this medium. The amount of data it contains, first of all. There has never been a corpus of language data as large as this one. It now contains more written language than all the libraries in the world combined, and the time it takes for its informational content to double in size is soon going to be measured in hours, as more parts of the world come online, and in seconds, as voice-over-internet becomes routine.

Secondly, there is the diversity of digitally mediated communication. The stylistic range of digitally mediated communication has to recognise not only internet texts, but also the vast outputs found in email, chatrooms, virtual worlds, blogging, instant messaging, and text messaging, as well as the increasing amount of linguistic communication in social networking forums (over 100 in 2008) such as Facebook, MySpace, Hi5, and Bebo. Each of these domains presents different communicative perspectives, properties, strategies, and expectations. It is impossible, at present, to find a linguistic generalization that applies comfortably to digitally mediated communication as a whole.

Part of the reason for this is another linguistically challenging property: the speed of change. It is difficult to keep pace with the communicative opportunities offered by new technologies, let alone to explore them in the required linguistic detail. By way of anecdotal illustration, the first edition of my *Language and the Internet* appeared in 2001: it made no reference to blogging and instant messaging, which had achieved no public presence at that time. A new edition of the book was therefore quickly needed, and that appeared in 2006. It included sections on the language of blogs and of instant messages, but it made no reference to the social networking sites, which were coming into prominence at the time. Linguistic studies of digitally mediated communication seem always to be out of date as soon as they appear.

Even within a single domain, it is difficult to keep pace. How can we generalize about the linguistic style of emails? When it first became prevalent, in the mid-90s, the average age of emailers was in the 20s. Today, it is in the late 30s: the average in the UK rose from 35.7 to 37.9 in the year October 2006–October 2007, according to Nielsen Online. Doubtless similar increases are to be found in other countries. This means that many emailers, for example, are now senior citizens. The consequence is that the original colloquial and radical style of emails (with its deviant spelling, punctuation, and capitalization) has been supplemented by a more conservative and formal style, as older people introduce their norms derived from the standard language. I now receive many emails which follow the conventions of a traditional letter, beginning 'Dear David', concluding 'Yours sincerely', and so on. This would have been unusual among the geeks who first explored the medium a few decades ago.

Another linguistically challenging property of digitally mediated communication is perhaps the most surprising: the inaccessibility of much of it. There is of course no problem in finding and downloading data from the pages of the Web, within the various legal and commercial constraints imposed by website owners. But it is a different matter when dealing with such domains as emailing, chatrooms, and texting. People are notoriously reluctant to allow their private e-communications to be accessed by passing linguists. There are now some corpora of emails and chatroom interaction, but issues of reliability and representativeness have yet to be fully explored, and some domains, such as text-messaging, remain elusive. The research literature is characterized by a great deal of theoretical speculation but relatively few empirical studies.

And a final example of the linguistic challenge. Assuming we have somehow achieved access to the data, how do we handle the unprecedented specificity of the linguistic information it contains? Linguists are used to being vague when it comes to describing language change: a word is said to have entered the language 'in the early sixteenth century' or in 'the 1780s'. Indeed, with rare exceptions, it has been impossible to identify the precise moment at which a new word or sense arrives in a language. But the time-stamping of webpages, and the ability to track changes, opens up a whole new set of opportunities. If I introduce a new word such as *digitextualization* on my website tomorrow at 9.42, it will be possible for lexicographers to say that the first recorded use of this word was at 9.42 am on Friday 30 October 2008. This sort of chronological specificity has hitherto been of professional interest only to forensic linguists, concerned to identify patterns of criminal interaction, but it will in future be of much broader relevance. However, I am currently unclear about how linguists will approach the handling of this level of descriptive detail.

New medium or old?

Notwithstanding the challenges, linguists have been able to make some progress in understanding the properties of digitally mediated communication. 'How do the new tools change our concepts and categories?' is one of the issues that are considered in the contributions to this book. 'Texts as objects of transmission' is another. I agree that the matter is best addressed initially through text comparison: comparing digitally mediated communication with the familiar notions of text as found in such domains of linguistic study as textlinguistics, discourse analysis, and stylistics. Three language modalities are traditionally recognized: speech, writing, and—not dealt with here—sign, in the sense of sign language used among deaf people. To what extent are these modalities replicated in digitally mediated communication? And is digitally mediated communication an imitation of the medium of the book? We must generalize these questions: 'Is digitally mediated communication an imitation of the medium of writing, whether in book or other form?' And we must furthermore also extend them: 'Is digitally mediated communication an imitation of the speech modality?' And of both we can ask the question originally referred to in the colloquium theme: 'Do we speed up classical techniques, or do we develop a new domain of techniques for access to [...] texts?' In short, are we doing things we could not do before? My answer, as we will see, is both 'yes and no'.

How are we to compare mediums of communication? The anthropological and zoological approaches to semiotics have shown us the fruitfulness of a design-feature framework, in which salient properties of communication are identified and used as a basis of comparison. In general linguistics, this procedure was first introduced by Charles Hockett in his comparison of language with animal communication.[1] In text linguistics, it takes two forms: an analysis of the formal properties of the mediums, and an analysis of their pragmatic properties; that is, looking at the intentions and effects relating to their use.

---

[1] C. Hockett, A course in modern linguistics.

Formal properties

The formal properties of speech and writing are displayed in Table 1 and compared with digitally mediated communication domains in

Table 1: Differences between speech and writing (after Crystal, 1995)

Speech	Writing
1. Time-bound Speech is time-bound, dynamic, transient. It is part of an interaction in which both participants are usually present, and the speaker has a particular addressee (or several addressees) in mind.	*1. Space-bound* Writing is space-bound, static, permanent. It is the result of a situation in which the writer is usually distant from the reader, and often does not know who the reader is going to be.
2. Spontaneous There is no time-lag between production and reception, unless one is deliberately introduced by the recipient. The spontaneity and speed of most speech exchanges make it difficult to engage in complex advance planning. The pressure to think while talking promotes looser construction, repetition, rephrasing, and comment clauses (for example *you know, you see*). Intonation and pause divide long utterances into manageable chunks, but sentence boundaries are often unclear.	*2. Contrived* There is always a time-lag between production and reception. Writers must anticipate its effects, as well as the problems posed by having their language read and interpreted by many recipients in diverse settings. Writing allows repeated reading and close analysis, and promotes the development of careful organization and compact expression, with often intricate sentence structure. Units of discourse (sentences, paragraphs) are usually easy to identify through punctuation and layout.
3. Face-to-face Because participants are typically face-to-face, they can rely on such extralinguistic cues as facial expression and gesture to aid meaning (feedback). The lexicon is often vague, using words which refer directly to the situation (deictic expressions, such as *that one, in here, right now*).	*3. Visually decontextualized* Lack of visual contact means that participants cannot rely on context to make their meaning clear; nor is there any immediate feedback. Most writing therefore avoids the use of deictic expressions, which are likely to be ambiguous.

Table 1 (*cont.*)

Speech	Writing
4. Loosely structured Many words and constructions are characteristic of (especially informal) speech, such as contracted forms (*isn't*). Lengthy coordinate sentences are normal, and are often of considerable complexity. There is nonsense vocabulary (for example *thingamajig*), obscenity, and slang, some of which may appear as graphic euphemism (*f****).	*4. Elaborately structured* Some words and constructions are characteristic of writing, such as multiple instances of subordination in the same sentence, elaborately balanced syntactic patterns, and the long (often multi-page) sentences found in some legal documents. Certain items of vocabulary are never spoken, such as the longer names of chemical compounds.
5. Socially interactive Speech is very suited to social or 'phatic' functions, such as passing the time of day, or any situation where casual and unplanned discourse is desirable. It is also good at expressing social relationships and personal attitudes, due to the vast range of nuances which can be expressed by the prosody and accompanying nonverbal features.	*5. Factually communicative* Writing is very suited to the recording of facts and the communication of ideas, and to tasks of memory and learning. Written records are easier to keep and scan, tables demonstrate relationships between things, notes and lists provide mnemonics, and text can be read at speeds which suit a person's ability to learn.
6. Immediately revisable There is an opportunity to rethink an utterance while the other person is listening (starting again, adding a qualification). However, errors, once spoken, cannot be withdrawn; the speaker must live with the consequences. Interruptions and overlapping speech are normal and highly audible.	*6. Repeatedly revisable* Errors and other perceived inadequacies in our writing can be eliminated in later drafts without the reader ever knowing they were there. Interruptions, if they have occurred while writing, are also invisible in the final product.
7. Prosodically rich Unique features of speech include most of the prosody. The many nuances of intonation, as well as contrasts of loudness, tempo, rhythm, pause, and other tones of voice cannot be written down with much efficiency.	*7. Graphically rich* Unique features of writing include pages, lines, capitalization, spatial organization, and several aspects of punctuation. Only a very few graphic conventions relate to prosody, such as question marks and italics. Several written genres (for example timetables, graphs) cannot be read aloud efficiently, but have to be assimilated visually.

Tables 2a and 2b. Disregarding the differences between situations in Tables 2a and 2b, and looking solely at the cells in terms of 'yes', 'variable', and 'no', it is plain that digitally mediated communication has far more properties linking it to writing than to speech. Of the 42 cells in the speech summary in Table 2a, only 15 are 'yes', 4 are 'variable', and 23 are 'no'. The situation for the writing summary in Table 2b, as we would expect, is almost exactly the inverse: 11 are 'yes', 8 are 'variable', and 23 are 'no'. On the whole, this comparison shows that digitally mediated communication is better seen as written language which has been pulled some way in the direction of speech than as spoken language which has been written down. However, expressing the question in terms of the traditional dichotomy is itself misleading. Digitally mediated communication is identical to neither speech nor writing, but selectively and adaptively displays properties of both. It is more than an aggregate of spoken and written features. It does things that neither of these other mediums do. There is nothing in digitally mediated communication like the phenomenon of simultaneous feedback ('mhm'-type vocalizations, nods, facial expressions from the listener while the speaker is talking) in face-to-face conversation;

Table 2a: Spoken language criteria applied to selected digitally mediated communication areas (Crystal, 2006)

		Web	Blogging	Email	Chatgroups	Virtual worlds	Instant messaging
1	time-bound	no	no	yes, but in different ways	yes, but in different ways	yes, but in different ways	yes
2	spontaneous	no	yes, but with restrictions	variable	yes, but with restrictions	yes, but with restrictions	yes
3	face-to-face	no	no	no	no	no	no, unless camera used
4	loosely structured	variable	yes	variable	yes	yes	yes
5	socially interactive	no, with increasing options	no, with increasing options	variable	yes, but with restrictions	yes, but with restrictions	yes
6	immediately revisable	no	no	no	no	no	no
7	prosodically rich	no	no	no	no	no	no

Table 2b: Written language criteria applied to selected digitally mediated communication areas (Crystal, 2006)

		Web	Blogging	Email	Chatgroups	Virtual worlds	Instant messaging
1	space-bound	yes, with extra options	yes	yes, but routinely deleted	yes, but with restrictions	yes, but with restrictions	yes, but moves off-screen rapidly
2	contrived	yes	variable	variable	no, but with some adaptation	no, but with some adaptation	no
3	visually decontextualized	yes, but with much adaptation	yes	yes	yes	yes, but with some adaptation	yes, unless camera used
4	elaborately structured	yes	variable	variable	no	no	no
5	factually communicative	yes	yes	yes	variable	yes, but with some adaptation	variable
6	repeatedly revisable	yes	variable	variable	no	no	no
7	graphically rich	yes, but in different ways	no, with increasing options	no	no	yes, but in different ways	no

and the hypertext link, which is the obligatory, fundamental functional unit of the Web, has no equivalent in traditional writing aside from such optional features as the footnote and cross-reference.

What makes digitally mediated communication so interesting is the way it relies on characteristics belonging to both sides of the speech/writing divide. At one extreme is the Web, which in many of its functions (for example databasing, reference publishing, archiving, advertising) is no different from traditional situations which use writing; indeed, most varieties of written language (legal, religious, and so on) can now be found on the Web with little stylistic change other than an adaptation to the electronic medium. In contrast, the situations of e-mail, chatgroups, virtual worlds, and instant messaging, though expressed through the medium of writing, display several of the core properties of speech. They are time-governed, expecting or demanding an immediate response; they are transient, in the sense that messages may be immediately deleted (as in emails) or be lost to attention as they scroll off the screen (as in chatgroups); and their utterances display much of the urgency and energetic force which is characteristic of face-to-face conversation. The situations are not all equally 'spoken' in character. We 'write' emails, not 'speak' them. But chatgroups are for 'chat', and people certainly 'speak' to each other there—as do people involved in virtual worlds and instant messaging. The interesting question is whether the technology makes us do this in different ways, and offers opportunities to do this in new ways.

Digitally mediated communication and speech

There are several differences between digitally mediated communication and speech that produce new properties in texts. Here are two of them.

Turn-taking

In a traditional speech setting, it is impossible to hold a conversation with more than one or two people at a time. Entering a room in which several conversations are taking place at the same time, we cannot pay attention to all of them or interact with all of them. But in multi-party settings in digitally mediated communication, this is perfectly feasible and normal. In a chatroom, we observe messages from other participants scrolling down the screen: there may be several conversations going on, on different topics, and we can attend to them all, and respond to them, depending only on our interest, motivation, and typing speed. If we were to attempt to interact in this way in a cocktail party, for example, we would be locked up!

Does this make us use language differently? Indeed it does, for it introduces a wholly new set of options for the *conversational turn*. Turn-taking is so fundamental to conversation that most people are not conscious of its significance as a means of enabling interactions to be successful. But it is a conversational fact of life that people follow the routine of taking turns, when they talk, and avoid talking at once or interrupting each other randomly or excessively. Moreover, they expect certain 'adjacency-pairs' to take place: questions to be followed by answers, and not the other way round; similarly, a piece of information to be followed by an acknowledgement, or a complaint to be followed by an excuse or apology. These elementary strategies, learned at a very early age, provide a normal conversation with its skeleton.

In digitally mediated communication, the turn-taking, as seen on a screen, is very different, because it is dictated by the software, and not by the participants. In a chatgroup, for instance, even if one did start to send a reaction to someone else's utterance before it was finished, the reaction would take its turn in a non-overlapping series of utterances on the screen, dependent only on the point at which the send signal was received at the host server. Messages are posted to a receiver's screen linearly, in the order in which they are received by the system. In a multi-user environment, messages are coming in from various sources all the time, and with different lags. Because of the way packets of information are sent electronically through different global routes, between sender and receiver, it is even possible for turn-taking reversals to take place, and all kinds of unpredictable overlaps to appear. Lucy asks a question; Sue receives it and sends an answer, but on Ben's screen the answer is received before the question. Or, Lucy sends a question, Sue replies, and Lucy sends another question; but on Ben's screen the second question arrives before Sue's reply to the first. The number of overlapping interactions that a screen may display at any one time increases depending on the number of participants and the random nature of the lags. What is surprising is that practised participants seem to tolerate (indeed revel in) the chaos which ensues. The issue is now receiving a great deal of empirical study.

Emoticons

Apart from in audio/video interactions (such as iChat), digitally mediated communication lacks the facial expressions, gestures, and conventions of body posture and distance (the *kinesics* and *proxemics*) which are so critical in expressing personal opinions and attitudes and in moderating social relationships. The limitation was noted early in the development of the medium, and led to the introduction of *smileys* or *emoticons*, such as the basic pairing of :) and : (for positive and negative reactions respectively. Today there are some 60 or so emoticons offered by some message exchange systems, though totals vary. It is plain that they are a potentially helpful way of capturing some of the basic features of facial expression, but their semantic role is limited. They can forestall a gross misperception of a speaker's intent, but an individual smiley still allows a huge number of readings (happiness, joke, sympathy, good mood, delight, amusement, etc.) which can only be disambiguated by referring to the verbal context. Without care, moreover, they can lead to their own misunderstanding: adding a smile to an utterance which is plainly angry can increase rather than decrease the force of the 'flame'. It is a common experience that a smile can go down the wrong way: 'And you can wipe that smile off your face, as well!'

What is interesting to the linguist, of course, is why these novelties have turned up now. Written language has always been ambiguous, in its omission of facial expression, and in its inability to express all the intonational and other prosodic features of speech. Why did no one ever introduce smileys there? The answer must be something to do with the immediacy of digitally mediated communication interaction, its closeness to speech. In traditional writing, there is time to develop phrasing which makes personal attitudes clear; that is why the formal conventions of letter-writing developed. And when they are missing, something needs to replace them. A rapidly constructed digitally mediated communication message, lacking the usual courtesies, can easily appear abrupt or rude. A smiley defuses the situation. We might therefore expect to see the frequency of smileys reduce, as people get more used to digitally mediated communication exchanges and construct their messages more carefully and explicitly. My impression is that they are at present most frequent in instant messaging, which is the most quickfire of all digitally mediated communication interactions.

Other points of difference

Absent also are other linguistic features typical of conversational speech, and these make it even more difficult for language to be used on the Internet in a truly conversational way. These limitations arise out of the current dependence of the medium on typing speed and ability. The fact of the matter is that even the fastest typist comes nowhere near the spontaneity and speed of speech, which in conversation routinely runs at five or six syllables a second. Even apparently spontaneous digitally mediated communication messages can involve elements of preplanning, pausing to think while writing, and mental checking before sending, which are simply not options in most everyday conversation. Some features of spoken language are often present in Internet writing, such as short constructions, phrasal repetition, and a looser sentence construction. But studies of e-mail and chatgroup interactions have shown that they generally lack the very features of spoken language which indicate most spontaneity—notably, the use of reaction signals (*m, mhm, uh-huh, yeah*...) and comment clauses (*you know, you see, mind you*...). Indeed, some writers have identified the lack of these features as one of the reasons why so many Internet interactions are misperceived as abrupt, cold, distant, or antagonistic. Addressing someone on the Internet is a bit like having a telephone conversation in which a listener is giving you no reactions at all: it is an uncomfortable and unnatural situation, and in the absence of such feedback one's own language becomes more awkward than it might otherwise be.

Digitally mediated communication and writing

If digitally mediated communication does not display the properties we would expect of speech, does it instead display the properties we expect of writing? Here too, there are important points of difference.

Persistence

Let us consider first the space-bound character of traditional writing— the fact that a piece of text is static and permanent on the page. If something is written down, repeated reference to it will be an encounter with an unchanged text. We would be astonished if, upon returning to a particular page, it had altered its graphic character in some way. Putting it like this, we can see immediately that digitally mediated communication is not by any means like conventional writing. A 'page' on the Web often varies from encounter to encounter (and all have the option of varying, even if page-owners choose not to take it) for several possible reasons: its factual content might have been updated, its advertising sponsor might have changed, or its graphic designer might have added new features. Nor is the writing that you see necessarily static, given the technical options available which allow text to move around the screen, disappear/reappear, change colour, and so on. From a user point of view, there are opportunities to 'interfere' with the text in all kinds of ways that are not possible in traditional writing. A page, once downloaded to the user's screen, may have its text cut, added to, revised, annotated, even totally restructured, in ways that nonetheless retain the character of the original. The possibilities are causing not a little anxiety among those concerned about issues of ownership, copyright, and forgery.

The other Internet situations also display differences from traditional writing, with respect to their space-bound presence. Emails are in principle static and permanent, but routine textual deletion is expected procedure (it is a prominent option in the management system), and it is possible to alter messages electronically with an ease and undetectability which is not possible when people try to alter a traditionally written text. Messages in asynchronic chatgroups and blogs tend to be long-term in character; but those in synchronic groups, virtual worlds, and instant messaging are not. In the literature on digitally mediated communication, reference is often made to the *persistence* of a conversational message—the fact that it stays on the screen for a period of time (before the arrival of other messages replaces it or makes it scroll out of sight). This certainly introduces certain properties to the conversation which are not available in speech. It means, for example, that someone who enters a conversation a couple of turns after an utterance has been made can still see the utterance, reflect upon it, and react to it; the persistence is relatively short-lived, however, compared with that routinely encountered in traditional writing. It also means, for those systems that provide an archiving log of all messages, in the order in which they were received by the server, that it is possible in principle to browse a past conversation, or search for a particular topic, in ways that spontaneous (unrecorded) conversation does not permit; however, in practice none of the systems currently available enable this to be done with ease, time-lags and the other factors described above making it extremely difficult to follow a topical thread in a recorded log. There are well-established means of finding one's way through a traditional written text: they are called indexes, and they are carefully compiled by indexers, who select and organize relevant information. Indexes of this kind are not likely in interactive digitally mediated communication, because there is so much of it and the subject-matter does not usually warrant it. There has been little research into the question of whether automatic indexing could be adapted so as to provide useful end-products.

Other points of difference

The other characteristics of traditional written language also display an uncertain relationship to digitally mediated communication. Is digitally mediated communication contrived, elaborate in its construction, and repeatedly revisable (items 2, 4, and 6 in Table 1)? For the Web, the answer has to be yes, allowing the same range of structural complexity as would be seen elsewhere. For chatgroups, virtual worlds, and instant messaging, where the pressure is strong to communicate rapidly, the answer has to be no, though the fact that smileys and other graphic conventions have been devised illustrates a certain degree of contrivance. Blogs vary greatly in their constructional complexity: some blogs are highly crafted; others are wildly erratic, when compared with the norms of the standard written language. (It should be borne in mind that blogging is the first continuous-text public written genre which has not been subjected to the moderating influence of editors, copy-editors, and proof-readers since the Middle Ages.) Emails vary enormously: some people are happy to send messages with no revision at all, not caring if typing errors, spelling mistakes, and other anomalies are included in their messages; others take as many pains to revise their messages as they would in non-digitally mediated communication settings.

Is digitally mediated communication visually decontextualized (item 3 in Table 2b)? Immediate visual feedback is always absent, as discussed above, so in this respect digitally mediated communication is just like traditional writing. But Web pages often provide visual aids to support text, in the form of photographs, maps, diagrams, animations, and the like; and many virtual-world settings have a visual component built in. The arrival of webcams is also altering the communicative dynamic of digitally mediated communication interactions, especially in instant messaging.

Is digitally mediated communication factually communicative (item 5 in Table 2b)? For the Web, blogs, and emails, the answer is a strong yes. The other two situations are less clear. Within the reality parameters established by a virtual world, factual information is certainly routinely transmitted, but there is a strong social element always present which greatly affects the kind of language used. Chatgroups vary enormously: the more academic and professional they are, the more likely they are to be factual in aim (though often not in achievement, if reports of the amount of flaming are to be believed); the more social and ludic chatgroups, on the other hand, routinely contain sequences which have negligible factual content. Instant message exchanges are also highly variable, sometimes containing a great deal of information, sometimes being wholly devoted to social chit-chat.

Finally, is digitally mediated communication graphically rich? Once again, for the Web the answer is yes, its richness having increased along with technological progress, putting into the hands of the ordinary user a range of typographic and colour variation that far exceeds the pen, the typewriter, and the early word processor, and allowing further options not available to conventional publishing, such as animated text, hypertext links, and multimedia support (sound, video, film). On the other hand, as typographers and graphic designers have repeatedly pointed out, just because a new visual language is available to everyone does not mean that everyone can use it well. Despite the provision of a wide range of guides to Internet design and desk-top publishing, examples of illegibility, visual confusion, over-ornamentation, and other inadequacies abound. They are compounded by the limitations of the medium, which cause no problem if respected, but which are often ignored, as when we encounter screenfuls of unbroken text, paragraphs which scroll downwards interminably, or text which scrolls awkwardly off the right-hand side of the screen. The difficulties are especially noticeable in blogging, where many pages fail to use the medium to best effect. The problems of *graphic translatability* are only beginning to be appreciated—that it is not possible to take a paper-based text and put it on a screen without rethinking the graphic presentation and even, sometimes, the content of the message. Add to all this the limitations of the technology. The time it takes to download pages which contain 'fancy graphics' and multimedia elements is a routine cause of frustration, and in interactive situations can exacerbate communicative lag.

Pragmatic properties

Pragmatics (within linguistics) studies the choices available to users when they speak or write, and the factors which govern their selection, such as the intention they have in mind or the effect they wish to convey. Some quite sophisticated classifications of texts have been made, which often provide a methodological framework for corpus construction. An example is the classification which informs the Survey of English Usage at University College London (Table 3). Each of the recognized categories has its own formal character.

From a pragmatic perspective, we would expect a new medium to motivate the appearance of different kinds of texts, reflecting the aims

Table 3: Pragmatic classification of mixed-medium texts (Crystal 1997, 292)

Speech
  To be heard
    Now (the norm)
    Later, for example telephone answering messages
  To be written down
    As if spoken, for example police statement, magazine interview
    As if written, for example letters, dictation

Writing
  To be read (the norm)
  To be read aloud
    As if spoken, for example radio/TV drama
    As if written, for example radio/TV newsreading
  To be partly read aloud, for example broadcasting continuity summaries

Mixed medium
  To self, for example memoranda, shopping list
  To single other, for example co-authorship sessions
  To many others, for example spoken commentary on a handout or
    blackboard

and intentions of the users. And so it proves to be. The content displayed on a screen permits a variety of textual spaces whose pragmatic purpose varies, some of which are summarized in Table 4. In the present state of research, a list of this kind can only be representative, not comprehensive. And the stylistic analysis of the texts relating to each of these categories is in its infancy. Plainly, there is a scale of online adaptability. At one extreme, we find texts where no adaptation to digitally mediated communication has been made—a pdf of an article on screen, for example, with no search or other facilities—in which case, any linguistic analysis would be identical with that of the corresponding offline text. At the other extreme, we find texts which have no counterpart in the offline world. Here are four examples.

Texts whose aim is to defeat spam filters

We only have to look in our email junk folder to discover a world of novel texts whose linguistic properties sometimes defy analysis.

    supr vi-agra online now znwygghsxp
    VI @ GRA 75% off regular xxp wybzz lusfg
    fully stocked online pharmac^y
    Great deals, prescription d[rugs

Table 4: Pragmatic classification of digitally mediated communication texts

digitally mediated communication text
As an end in itself
To be read (the norm)
In the displayed language
In another language (if available), for example translate, click icon
To be read
Statically (the norm)
Dynamically, for example news feeds, incoming results, market reports
To be added to, for example chatroom, forum, post a comment
To be acted upon
To obtain information, for example contact us, help
To review or evaluate, for example consumer reviews
To persuade, for example ads, wish lists, more like this
To purchase, for example payment methods, buying procedures
As a means to an end
On the same page
To be searched, for example advanced search, archive, track order
On a different page
Hyperlinks, for example quick links, permalinks
Using another medium, for example podcast, video link

It is possible to see a linguistic rationale in the graphological variations in the word *Viagra*, for example, introduced to ensure that it avoids the word-matching function in a filter. We may find the letters spaced (*V i a g r a*), transposed (*Viarga*), duplicated (*Viaggra*), or separated by arbitrary symbols (*Vi*agra*). There are only so many options, and these can to a large extent be predicted (an issue familiar to cryptologists). There have been huge advances here since the early days when the stupid software, having been told to ban anything containing the string S-E-X, disallowed messages about Sussex, Essex, and many another innocent term. There is also an anti-linguistic rationale, as one might put it, in which random strings are generated (*wybzz*). These too can be handled, if one's spam filter is sophisticated, by telling it to remove any message which does not respect the graphotactic norms of a language (i.e. the rules governing syllable structure, vowel sequence, and consonant clusters).

Texts whose aim is to guarantee higher rankings in web searches

How is one to ensure that one's webpage appears in the first few hits in a web search? There are several techniques, some nonlinguistic, some linguistic. An example of a nonlinguistic technique is the frequency of hypertext links: the more pages link to my site, the more likely my page will move up the rankings. An example of a linguistic technique is the listing of key words or phrases which identify the semantic content of a page in the page's metadata: these will be picked up by the search engine and given priority in a search. Neither of these techniques actually alters the linguistic character of the text on a page. Rather different is a third technique, where the text is manipulated to include keywords, especially in the heading and first paragraph, to ensure that a salient term is prioritized. The semantic difference can be seen in the following pair of texts (invented, but based on exactly what happens). Text A is an original paragraph; text B is the paragraph rewritten with ranking in mind, to ensure that the product name gets noticed:

(A) The Crystal Knitting-Machine is the latest and most exciting product from Crystal Industries. It has an aluminium frame, comes in five exciting colours, and a wide range of accessories.
     The Crystal Knitting-Machine is the latest and most exciting product from Crystal Industries.
(B) • The Crystal Knitting-Machine has an aluminium frame.
     • The Crystal Knitting-Machine comes in five exciting colours.
     • The Crystal Knitting-Machine has a wide range of accessories.

Some search engines have now got wise to this technique, and are trying to block it, but it is difficult, in view of the various paraphrases which can be introduced (for example *Knitting-Machine from Crystal, Crystal Machines for Knitting*).

Texts whose aim is to save time, energy, or money

Text-messaging is a good example of a text genre whose linguistic characteristics have evolved partly as a response to technological limitations. The limitation to 160 characters (for Roman alphabets) has motivated an increased use of nonstandard words (of the *C U l8r* type), using logograms, initialisms, shortenings, and other abbreviatory conventions. The important word is 'partly'. Most of these abbreviations were being used in digitally mediated communication long before mobile phones became a routine part of our lives. And the motivation to use them goes well beyond the ergonomic, as their playful character provides entertainment value as an end in itself as well as increasing rapport between participants.[2]

Another example of a new type of text arising out of considerations of convenience is the email which uses *framing*. We receive a message which contains, say, three different points in a single paragraph. We can, if we want, reply to each of these points by taking the paragraph, splitting it up into three parts, and then responding to each part separately, so that the message we send back then looks a bit like a play dialogue. Then, our sender can do the same thing to our responses, and when we get the message back, we see his replies to our replies. We can then send the lot on to someone else for further comments, and when it comes back, there are now three voices framed on the screen. And so it can go on—replies within replies within replies—and all unified within the same screen typography. People find this method of response extremely convenient—to an extent, for there comes a point where the nested messages make the text too complex to be easily followed. I have never seen an e-exchange which goes beyond six levels of nesting.

Related to framing is intercalated response. Someone sends me a set of questions, or makes a set of critical points about something I have written. I respond to these by intercalating my responses between the points made by the sender. For clarity, I might put my responses in a different colour, or include them in angle brackets or some such convention. A further response from the sender might lead to the use of an additional colour; and if other people are copied in to the exchange, some graphical means of this kind, to distinguish the various participants, is essential.

Texts whose aim is to maintain a standard

Although the Internet is supposedly a medium where freedom of speech is axiomatic, controls and constraints are commonplace to avoid abuses. These range from the excising of obscene and aggressive language to the editing of pages or posts to ensure that they stay focused on a particular topic. Moderators (facilitators, managers, wizards; the terminology is various) have to deal with organizational,

---

[2] D. Crystal, *Txtng*.

social, and content-related issues.[3] From a textual point of view, what we end up with is a sanitized text, in which certain parts of language (chiefly vocabulary) are excluded, thus making it a species of restricted language. It is not clear how far such controls will evolve, as the notion of textual responsibility relating to the libel laws is still in the process of being tested.[4]

A good example of content moderation is in the online advertising industry, where there is a great deal of current concern to ensure that ads on a particular web page are both relevant and sensitive to the content of that page. Irrelevance or insensitivity leads to lost commercial opportunities and can generate extremely bad PR. Irrelevance can be illustrated by the CNN report of a street stabbing in Chicago, where the ads down the side of the screen said such things as 'Buy your knives here'—the software being unaware that the weapons sense of 'knife' in the news report did not match the cutlery sense of 'knife' in the ad inventory. Insensitivity can be illustrated by a page which was describing heritage visits to Auschwitz; the same silly software, having found 'gas' mentioned several times on the page, linked this with a power company's ads for 'cheap gas', much to the embarrassment of all concerned. Putting this in the terminology of pragmatics, the perlocutionary effect was not what was intended. The solution known as 'semantic targetting', as used in Ad Pepper Media's iSense and Sitescreen products, carries out a complete lexical analysis of web pages and ad inventories so that subject-matter is matched and ad misplacements avoided.[5] In extreme cases, such as a firm which does not want its ad to appear on a particular page (for example a child clothing manufacturer on an adult porn site), ads can be blocked from appearing. As a result, from a content point of view, the text that appears on a page appears more semantically coherent and pragmatically acceptable than would otherwise be the case.

Authorship issues

Framed, intercalated, and moderated texts illustrate a multi-authorship phenomenon which reaches its extreme in wiki-type texts, where

---

[3] M. Collins and Z.L. Berge, *The Moderators Home Page*.
[4] Concurring Opinions.
[5] Ad Pepper Media, 'iSense'.

people may alter an existing text as their inclination takes them. Social and legal issues aside, what does this phenomenon do to the texts themselves?

First of all, it makes texts stylistically heterogeneous, as a glance at most Wikipedia pages will show. Sometimes there are huge differences, with standard and nonstandard language coexisting on the same page, often because some of the contributors are plainly communicating in a second language in which they are nonfluent. Traditional notions of stylistic coherence, with respect to level of formality, technicality, and individuality, no longer apply, though a certain amount of accommodation is apparent, with contributors sensing the properties of each other's style. Secondly, it makes texts pragmatically heterogeneous, as the intentions behind the various contributions vary greatly. Wiki articles on sensitive topics illustrate this most clearly, with judicious observations competing with contributions that range from mild through moderate to severe in the subjectivity of their opinions.

Thirdly, and fundamentally, it disturbs our sense of the physical identity of a text. How are we to define the boundaries of a text which is ongoing? People can now routinely add to a text posted online, either short-term (as in the immediate response to a news story), or medium- or long-term, as in comments posted to a blog, bulletin board, or other forum. Ferdinand de Saussure's classical distinction between synchronic and diachronic does not adapt well to digitally mediated communication, where everything is diachronic, time-stampable to a micro-level. Texts are classically treated as synchronic entities, by which we mean we disregard the changes that were made during the process of composition and treat the finished product as if time did not exist. But with many digitally mediated communication texts there is no finished product. I can today post a message to a forum discussion on page X from 2004. From a linguistic point of view, we cannot say that we now have a new synchronic iteration of X, because the language has changed in the interim. I might comment that the discussion reads like something 'out of Facebook'—which is a comment that could be made only after 2005, when that network began. I do not know how to handle this.

The problem exists even when the person introducing the various changes is the same. The author of the original text may change it—altering a Web page, or revising a blog posting. How are we to view the relationship between the various versions? The question is particularly relevant now that print-on-demand texts are becoming common. It is possible for me to publish a book very quickly and cheaply, printing only a handful of copies. Having produced my first print-run, I then decide to print another, but make a few changes to the file before I send it to the POD company. In theory (and probably increasingly common in practice), I can print just one copy, make some changes, then print another copy, make some more changes, and so on. The situation is beginning to resemble medieval scribal practice, where no two manuscripts were identical, or the typesetting variations between copies of Shakespeare's First Folio. The traditional terminology of 'first edition', 'second edition', 'first edition with corrections', ISBN numbering, and so on, seems totally inadequate to account for the variability we now encounter. But I do not know what to put in its place. The same problem is also present in archiving. The British Library, for example, has recently launched its Web Archiving Consortium.[6] My website is included. But how do we define the relationship between the various time-stamped iterations of this site, as they accumulate in the archive?

Anonymity issues

Digitally mediated communication is not the first medium to allow interaction between individuals who wish to remain anonymous, of course, as we know from the history of telephone and amateur radio; but it is certainly unprecedented in the scale and range of situations in which people can hide their identity, especially in chatgroups, blogging, and social networking. These situations routinely contain individuals who are talking to each other under nicknames (*nicks*), which may be an assumed first-name, a fantasy description (*topdude, sexstar*), or a mythical character or role (*rockman, elfslayer*). Operating behind a false persona seems to make people less inhibited: they may feel emboldened to talk more and in different ways from their real-world linguistic repertoire. They must also expect to receive messages from others who are likewise less inhibited, and be prepared for negative outcomes. There are obviously inherent risks in talking to someone we do not know, and instances of harassment, insulting or aggressive language, and subterfuge are legion. Terminology has evolved to

---

[6] The British Library, *Web-archiving Consortium.*

identify them, such as flaming, spoofing, trolling, and lurking.[7] New conventions have evolved, such as the use of CAPITALS to express 'shouting'. While all of these phenomena have a history in traditional mediums, digitally mediated communication makes them present in the public domain to an extent that was not encountered before. But we do not yet have detailed linguistic accounts of the consequences of anonymity.

Envoi

This paper is ending in the time-honoured scholarly way, of raising more questions than providing answers. But at this stage in the evolution of digitally mediated communication we have very little choice in the matter. Few of the communicative conventions of digitally mediated communication have received serious empirical linguistic study. The classical conception of a text is a selection of language by a known author directed at a known audience, expressing an intention which is specifiable, using a style which is coherent, and presented through a medium which is determinate in form. All of these attributes are at times uncertain, in digitally mediated communication. That is why we need this book.

References

Ad Pepper Media, 'iSense. The Semantic Revolution' (2008), http://www.isense.net/index.php?id=50&L=50.
Baron, N., *Always on. Language in an online and mobile world* (Oxford: OUP, 2008).
British Library, Web-archiving Consortium (2008), http://www.webarchive.org.uk.
Collins, M., and Z.L. Berge, The Moderators Home Page (2006), http://www.emoderators.com/moderators.shtml.
Concurring Opinions (2005), http://www.concurringopinions.com/archives/2005/11/suing_wikipedia_1.html.
Crystal, D., *The Cambridge encyclopedia of the English language*, 2nd edn. (Cambridge: CUP, 2003; first edn 1995).
——, *The Cambridge encyclopedia of language*, 2nd edn. (Cambridge: CUP, 1997).
——, *Language and the Internet*, 2nd edn. (Cambridge: CUP, 2006).
——, *Txtng. The Gr8 Db8* (Oxford: OUP, 2008).
Hockett, C.F., *A course in modern linguistics* (New York: Macmillan, 1958).
Neilson (2008), http://www.netratings.com/pr/pr_071218_UK.pdf.

---

[7] D. Crystal, *Language and the Internet*.

NEW MEDIUMS:
NEW PERSPECTIVES ON KNOWLEDGE PRODUCTION

Adriaan van der Weel

Print still being the primary medium for the exchange of knowledge today, we may be said to be homo typographicus, *living in the 'Order of the Book'. However, we are rapidly making a transition from analogue to digital textual transmission. As we make that transition we need to avoid the pitfall of imitating familiar aspects of existing mediums, stopping us from recognising and making full use of salient characteristics of the unfamiliar new medium. Imitation happens on the level of technological features, but also—more insidiously—on a conceptual level, and it is helpful to scrutinise the technological features and concepts concerned closely. All technologies, writing and print no less than the digital medium, result from a process of discovery of their social uses more than they are technological inventions. Not only is the digital medium too new to have reached the transparency of print yet, it is likely that this will never happen. For, the computer being a Universal Machine, its potential is unlimited, and the digital medium will continue to grow and expand along with it. That is not to say that it has no distinctive features at all; just that these are not immanent. The Universal Machine has an unlimited potential for accommodating different work practices, but only if we can articulate our demands, as producers, distributors, or end users, will we be able to influence the digital medium's development and harness its potential.*

For centuries the book, the journal and other forms of print have served scholars as their chief knowledge instruments. Outside of the academy, too, the book has had a central role in the transmission of knowledge. Take textbooks in education, or reports in government—all products of the printing press. This is of course rapidly changing today. Scholars are busy exploring the most diverse forms of digital communication, of which many impressive examples can already be found.[1] And again, the same is happening outside the academy. In

---

[1] Many were shown at the Colloquium on Text Comparison and Digital Creativity.

education electronic learning environments are taking over many traditional book functions; governments make use of many different forms of web-based communications, etcetera.

However, I would like to maintain that even today, amid the turbulent speed at which digital communications are usurping roles previously played by the book, most of our information habits remain book based. The extent to which we are still governed by the book and its institutions I like to indicate by the term 'Order of the Book'.[2] This Order of the Book is firmly based on a bibliographic frame of mind, fed by pervasive and intricate book-based information ordering, referencing and other aspects of our 'knowledge system'. But the persistent status of writing and print is equally apparent from the ubiquitous book-based metaphors that continue to govern our understanding of the broad spectrum of digital communications.

In the context of the current rapid transformation, it is particularly instructive to take a look at the dynamic history of the textual mediums. Past textual practices—in manuscript and print—can help equip us for a better understanding of the digital present.

Technological invention and social discovery

Rather than springing forth fully armed from the brain of a mythical inventor, the printed book as we know it today has a long history of gestation. By that I don't wish to suggest that Gutenberg should not be called an inventor. Even if his first attempts at printing with moveable type may well have been inspired by existing Korean and Chinese practices,[3] he did experiment with ways to cast the type, and introduced the art in Europe. However, the achievement of a technological *invention* is invariably matched by an often lengthy process of *discovery*. Discovery is a social development, involving crucially the recognition of an invention's social uses and significance.[4] The process of discovery typically takes a great deal longer than the act of inven-

---

[2] I explain the term more fully in my forthcoming book, provisionally entitled 'Digital text and the Order of the Book'.

[3] See Needham, *Science and civilisation in China*, 313–19.

[4] Scholars of Science and Technology Studies would refer to this discovery process as the social construction of social use and significance. Not only a technology is shaped over time, but also the market for it and labour and economic processes around it. Cf., for example, Wiebe Bijker, *Bicycles, bakelites and bulbs*.

tion, and it is a great deal more diffuse. For a shorthand illustration of this discovery process we can take a look at some aspects of two earlier textual mediums: printing and writing.

In the case of the printing press, Gutenberg may or may not have invented the technique of casting types for reproducing text on a printing press, but what he certainly did not invent was the reading machine that the book as a medium has since become. It took centuries of patient discovery before the printed book's distinctive characteristics were fully recognised. One particularly interesting example of such a characteristic is pagination. Page numbers depend on the property of identicalness of printed copies. While identicalness across copies of the *text* itself was certainly intended, in a bid to counter the unrelenting tendency towards corruption of manuscript transmission, identicalness of the text *on the individual page* was a coincidental side effect. Its potential as a helpful attribute in parsing the book's contents was only realised over time.[5] After Gutenberg's initial invention in the 1440s it was not till around 1475 that Nicolas Goetz of Cologne printed an edition of Werner Rolewinck's *Fasciculus temporum* using page numbers,[6] and it was not until almost a full century after Gutenberg's first printings that page numbering started to become common, exploiting an unintended but remarkable property of printing: that all its products are for most intents and purposes identical in content and form. Similarly it took time for an awareness to grow that printing, rather than being the coffin for the spoken and written word that John Donne still saw in it as late as the seventeenth century, was in fact able to safeguard texts, increasing their chances to stay alive by the factor of their print run. All in all it took centuries for print to become the fully developed and almost fully transparent medium it is in today's society, with all the subtly interacting parts that contribute to make it into such a well-oiled machine. Besides page numbers these include also, for example, the running head, the table of contents, the index, and all the other ingredients of its familiar interface.

---

[5] There is evidence that the popularity of page numbering in manuscripts was growing from about 1300, while numbering in incunables is extremely rare (just over 10%). Page numbering in (unique) manuscripts would obviously serve a different purpose than page numbering in printed books, and so there would be no *prima facie* case for continuity between the two practices. In fact, numbering in printed books may have begun as an aid to printers rather than readers. For a detailed discussion see Margaret M. Smith, 'Printed foliation'.

[6] Printed numbering of leaves had already occurred in 1470 (*ibid.*, 54).

Going even further back in history, writing went through a similar social development. As long as the number of initiates remained limited, no real social familiarity with writing could develop, and even a certain distrust of the written word could prevail. Comparing writing to the spoken word, Plato famously lamented the helpless fixity and unchangeability of the written text, which he thought left it unable to look after itself. He was also dismayed by its dangerous promiscuity. Once set down on paper, the author lost control over his words entirely, and a text could commune with anyone who encountered it.[7] Plato seems to attribute an agentive role to writing, as if it were less a dead object (as we are inclined to see in it) than an undead one. He certainly did not regard writing as alive: it lacked the presence of the spoken word.[8] He was not alone in this. Writers like Sappho, and later Horace stressed the deadness of the book's materiality and rated its longevity less than that of the living (i.e., spoken) word. At any rate, what Plato, Sappho and Horace shared is the suspicion that this technology was not socially very beneficial. Their reserve towards writing shows that various aspects of the materiality of writing that we now celebrate as milestones of human achievement took time to emerge and become dominant. One of these was its potential for promoting objectivity—both in the literal sense and in the metaphoric sense so coveted by Plato. Ironically this objectivity was something Plato was particularly keen to advocate.[9]

Even the purely mechanical development of writing had its social effects. The increasing application of word spacing from the eleventh and twelfth centuries, for example, represented a major step in the creation of a literate society.[10] Before writing became common, it was often 'speech set down', and thus without word spacing. Without the application of consistent word spacing from the eleventh century the great medieval educational reforms could not have brought about

---

[7] Plato's fears have been expertly analysed by John Durham Peters in *Speaking into the air*, 36–51.

[8] In fact, all writing, as all medial transmission, could be said similarly to lack the presence of the unmediated word—unless, that is, speech should also be regarded as a medium. However, I suggest that, being a function of the human body that does not require any technological 'extension', it should not.

[9] In *Preface to Plato* Eric Havelock suggests that it was the perceived subjectivity of poetry—resulting from its being transmitted orally and being learned by rote—that drove Plato to banning poetry from his ideal state.

[10] See Saenger, *Space between words*.

writing's new, more central place in society. It made reading technically easier, allowing more people beyond a small priviliged group of
initiates to read.

We are now witnessing the incunable period of the digital medium.
Despite being in its infancy, and even if the time frame is vastly shorter
than in the case of writing and printing, the digital medium as we
know it and use it today also has an—as yet brief—history of being
discovered as well as invented. Email, chatting, and word processing are among the most widely used applications on today's personal
computers. Though Charles Babbage and Ada Lovelace already realised in the middle of the nineteenth century that computers could
be programmed to calculate other things besides numbers, in actual
fact it took a long time for the computer to evolve from a calculating
machine to the language machine it has since become. Once computers could deal with text, in the shape of first word processing and then
desktop publishing, they were initially used chiefly in the service of the
existing textual medium: to replace typewriters, and to aid conventional print production.

There is no doubt that it was a triumph that we managed this replication of analogue textual practices and print functions in the digital
realm so well, from writing and editing to the origination of printing
plates, and more. Despite the triumph of the achievement it might be
argued, however, that such 'imitation'[11] of the analogue in the digital
did not really amount to what we now regard as the digital textual
medium. To the extent that this new medium replicated the known
functions of the book in the digital realm there was simply functional
continuity. But the digital textual medium as we now know it went
beyond mere replication. It could do something that none of the earlier textual mediums could do, taking digital textuality far beyond
that of the book. Again it took a process of *discovery* for this novel
medial potential to come into its own. Just as the technology of the
printing press was able to do something that manuscript couldn't, viz.
multiplication, the World Wide Web as a medium brought the new
dimension of distribution. Email had already been established for the
communication between remote terminals and the (mainframe) computer over the network. But again it took time and a major conceptual

---

[11] For want of a conscious motive, this cannot be constructed as 'imitation' in the
strict sense. What happens is that we cannot help but think in deep grooves.

leap before the computer's networking capabilities were fully recognised. With Tim Berners Lee's HyperText Markup Language in 1991 the internet was turned into a full-blown textual medium, the World Wide Web, the latest addition to the sequence of clay tablet, scroll, codex, printed book.

Each step in the process by which the computer turned into a digital textual medium has necessarily been based on the nature of the computer as a Universal Machine. Computers can be made to perform any task that may be expressed by way of an algorithm, including the algorithmic handling of letters, words, text and increasingly even spoken language. This means of course that the computer's medial functions can be infinitely extended beyond the textual and communicative ones we have already invented and/or discovered and are currently familiar with. Just as importantly, the computer's medial function is but one of an infinite array of other capabilities. This is especially relevant in the present scholarly context. As scholars we use computers not just— and not necessarily even primarily—as a communication medium to represent and disseminate our research findings, but also as a means to help us perform the research itself.

The sociotechnical nature of the digital textual medium

In what follows I would like to see what we can learn about the *nature* of the digital textual medium, with a particular view to explore the relationship between its technological properties and their social effects. For this purpose it will be useful, in conformance to the conceptual distinctions with which we are familiar as denizens of the Order of the Book, to try and make a distinction between the digital medium— roughly speaking, book functions gone digital—and the overarching 'digital textuality' which has enabled the digital medium to come into being in the first place. Digital textuality represents a much larger set of digitial textual practices, including, for example, the field that has become known as 'digital humanities' or 'humanities computing'.

If to understand digital textuality we look for inherent technological properties (comparable to such a property as the identity of reproduction in the printing press), the first and most fundamental technological property is that digital textuality is based on the Universal Machine. This can be well illustrated by the intrinsical differences between the process of gestation of the medium of the printed book on the one hand, and that of the digital medium on the other. I'll name two.

First, the book's technological properties, as distinct from the socio-technical uses we put them to, never really changed over the course of centuries. For example, we never managed to teach the printing press new tricks, such as, say, the multiplication of sound or moving images, or taking care of the distribution of the books it printed. The digital medium, by contrast, was a product of the unceasing development of the computer as a Universal Machine. In other words, while the book's gestation—the way it became a reading machine—was a process of discovery more than one of invention (the improvements after Gutenberg's first breakthrough occurred piecemeal over many centuries), digital textuality is a case of a continuous—and fundamentally unlimited—burst of inventions, each followed by its own process of discovery.

This has made for extremely fast development of digital textuality over the past half century. By all indicators (including, for instance, Moore's Law, that states that the number of transistors on a chip will double about every two years) that speed is not likely to abate. Moreover, digital textuality can be regarded as a hybrid phenomenon. On the one hand we approach it as a replacement for writing and printing; that is to say, as a medium with expectations shaped by the Order of the Book. These expectations are nicely visualised in Robert Darnton's famous 'communications circuit'.[12] In addition to the medial function expressed in the communications circuit, however, what digital textuality has to offer is the entire gamut of the programming capabilities of the Universal Machine. These take it far beyond the medial functions of the book and make it into an extraordinarily versatile research instrument—or, better, sociotechnical research environment. The medial function is, moreover, so seamlessly integrated in this research environment as to be virtually indistinguishable from it. That is to say that digital text can be dealt with by any imaginable application, whether it already exists, for example in other disciplines—including science—or is yet to be created.

The second intrinsic difference between the processes of gestation of the medium of the printed book and that of the digital textual medium is that the book has over the centuries gradually become more and more transparent as a technology. This was possible because the technological properties of the printing press and its products never fundamentally changed. Its 'user interface' is now so familiar that our

---

[12] Darnton, 'What is the history of books'.

expectations of its functionality are rarely thwarted. We know very well what it can and cannot do. Digital textuality, by contrast, is not so transparent. This realisation, viz. that transparency of the digital medium, however desirable, is not forthcoming, is problematic for us as scholarly media practitioners, and led to the Text Comparison and Digital Creativity colloquium. Our primary aim as scholars is to communicate our research findings successfully. Digital textuality should not stand in the way of that communication, or interfere with our research findings in any way. For that reason we would obviously want our digital textuality to be just as transparent as the products of print have over the centuries become. Our primary interest lies in scholarly communication, not in the intervening media technologies which make communication possible.[13]

I fear, though, that such transparency is fundamentally impossible. It is questionable whether we will ever be able to internalise the properties of digital textuality, or even of the digital textual medium. The reason for this is its protean nature. Digital textuality is not only fundamentally new and recent, having developed too rapidly for any comparable familiarity to evolve.[14] The fact that it is based on the programmable Universal Machine precludes such familiarity. In any case, the book's apparent transparency is no more than a chimera. It lulls us into a false sense that as a medium it is an undistorting clear window. Mediums are never value-free. They are not clear windows. Every medium represents a particular way to construct our knowledge. Notwithstanding the degree of familiarity we have reached with it, this applies also to the book. The difference is that we have internalised the properties of the book as a knowledge machine in a way we have not been able to do in the case of digital textuality, and most likely never will.

I fear that there is no simple way out of this predicament. The only realistic solution seems to be to become accustomed to the digital medium's changeability and to make ourselves thoroughly aware of what it does to our communication.

---

[13] What makes it possible to communicate *is* of course at the core of my own interests as a book historian—or, more properly, as a historian of textual transmission.

[14] Perhaps it has become transparent insofar as it may be possible to recognise a one-to-one relationship with the print media with which we were already so familiar.

Medial transformativity

This brings me to the notion of medial transformativity. We not only have to live with the lack of transparency and continuous change as innovation will continue at a high pace, but also with the attendant and ongoing process of discovering the social consequences of every new development. Despite the continuity in function (i.e., the transmission of text), there are in fact, as I have suggested, a great many discontinuities between the textual mediums. Each medium has its own bias, based on its technological properties. A simple example of what I mean is that the book and the digital textual medium represent two very different author–reader relationships. The book is a paper monument, which offers a fairly straightforward and widely recognised one-way hierarchical relationship between author and reader. It is not that readers don't bring meaning to the text; just that such creation of meaning is an individual and private affair which does not affect someone else's reading experience. Contrasting this with the more tentative and changeable, not to say fickle, form of the networked digital text brings into sharp focus its bidirectional and much more democratic nature. In these technological conditions, the shaping of meaning becomes a much more public and collaborative act.

At a higher level of abstraction, this medial bias may be constructed as a form of inevitable interpretation of the data to be transmitted. The chosen medial form dictates to a considerable extent what aspects of the data (say, someone's knowledge about an aspect of the world) are foregrounded, and what aspects will remain underexposed. 'Text is knowledge represented as matter', as the description of the theme session 'Knowledge creation and representation' in the invitation to the colloquium had it. Moving information from one medial carrier to another without changing it is impossible: each carrier will provide its own interpretation of that information.

In this sociotechnological process of medial transformativity, more often than not technology has the upper hand. We tend to invent technologies for which we only discover social uses afterwards rather than the other way around. Looking at manuscript and print we already considered some examples earlier, but the same goes for Edison's sound recording device, which he had intended for office dictation, or email, which developed out of the need to communicate with fellow mainframe users at a distance. No one could have foreseen the vast growth of the twentieth-centure music industry, or the central importance of email in every scholar's daily work, just as the mobile phone was not invented to extend total control over the life of everyone over the age of ten.

Stages of digital textual transmission

I have so far concentrated on the medial aspect of digital textuality and its history. With the help of the distinct stages I have identified in the history of computing as a technology and as a digital textual medium we can retrospectively recognise three relatively distinct phases also in social digital textual practice: the way we have as humanities researchers used the computer for text processing (in the widest sense).

While bearing in mind the remark made earlier about the unconscious nature of much of the 'imitation', *Stage 1* might be characterised as the imitation stage: doing the same things as we already did using analogue techniques, but with help from the machine.[15] It is a fascinating question to what extent searching, sorting, matching and the prodigious memory that computers have to offer go beyond being mere digital equivalents of old-fashioned analogue processes. Can the computer be programmed to do things that are fundamentally different, and not just faster, more precise, etcetera?

In this stage the computer was used in the service of traditional forms of knowledge creation. Both word processing and desktop publishing were geared towards the production of conventional 'hard copy'. That is to say, they were used to create traditional monuments attributed to named individuals, in the conventional paradigm of a hierarchical author–reader relationship. Father Busa's concordance to the work of Thomas Aquinas, begun in 1947, has been hailed as the world's first humanities computing project. However magisterial the project's scope and execution, it may be argued that it was ultimately in a conventional mould. While the *method* was new, the *result* was not, and it did not lead to new research questions.

In *Stage 2*, which includes the start of the World Wide Web, we witness the construction of a new medium for knowledge creation and, most importantly, dissemination. Crucially, it saw the birth and

---

[15] This 'imitation' stage is, incidentally, also a necessary stage in the sense that culture demands continuity. Memetic thinking corroborates this: only if ideas are accepted will they be successfully replicated (see Susan Blackmore, *The meme machine*).

development of the concept of markup, which was one of the chief technologies to further the dissemination function. Internet technology (based on the bidirectional flow of traffic) stimulated peer-to-peer horizontal collaboration. Yet at the same time the traditional hierarchical paradigm of knowledge production persisted, with experts producing knowledge to be consumed by end users.

A curious phenomenon that we can recognise in retrospect as characterising this stage was the emphasis on the production of raw data. The computer was very often used for the production of scholarly 'semi-manufactures'. This had no doubt much to do with the sheer magnitude of the investment of intellectual and financial resources in mastering digital work methods. One notable effect of this phenomenon was the deferral of the interpretive burden, which shifted more and more from the instigator of the scholarly communication to its recipient. For much the same reason, collaboration, both disciplinary, among humanities specialists, and interdisciplinary, between humanities specialists and computer scientists, was another notable characteristic of many humanities digitisation projects.

The most notable application of the concept of markup for use in the humanities still remains the Text Encoding Initiative (TEI). The history of its guidelines during this period, from 'P1'[16] down to 'P5',[17] testifies to the strength of its concept. The TEI has been able to remain at the forefront of textual work in the digital humanities for more than twenty years since its inception in 1988. Yet its strength continues to lie in monumental productions in the conventional mould (whether in print or digital), with the markup concentrating on forms of fixity inscribed by experts, whether they operate singly or collaboratively.

More recently we can be said to be entering *Stage 3*, the democratisation stage, which has also brought us Web 2.0. This marks the end of the 'hieratic period', with the scholarly and IT priesthood in charge. It is also characterised by the realisation that the computer offers more advanced processing abilities than we had been using in the humanities.

The reason we organise meetings like the Colloquium on Text Comparison and Digital Creativity (and they happen everywhere now) is

---

[16] C.M. Sperberg-McQueen and L. Burnard, eds, *Guidelines for the encoding and interchange of machine-readable texts.*
[17] *TEI P5: Guidelines for electronic text encoding and interchange.*

probably twofold. Firstly, we are becoming more aware of the fact that the computer represents a truly new medium. Though we have only just started our investigation of the technological properties of the digital textual medium, we are already beginning to find out about some of the particular ways in which it is so very different from print and, more importantly about the implications of those differences.[18] Secondly, there is a sense that we are staying behind compared to the sciences.[19] Whether justified or not, this fear has led to an increasing sense of urgency about moving humanities work practices into the digital realm. It is this third stage that I would now like to examine in a little more detail.

Building new knowledge instruments

With the growing awareness, not only of the fact that the digital medium is fundamentally different from the print medium, but also of the nature of the differences, we are in a position to think more consciously about the creation of new knowledge instruments to supplement—or replace—the printed book. It might be helpful to think of a number of basic categories of models in which the computer may be used in the creation of such new knowledge instruments.

Model A we might call the 'markup model'. It entails the formal inscription of pre-existing qualitative knowledge[20] in much the same way as has been done in manuscript and print for centuries. The chief difference is that the exigencies of the software (especially the rules which the markup has to obey in order to be 'valid') will frequently force previously implicit, unconscious knowledge to be made explicit. In addition, the markup model has evolved to accommodate the more 'democratic' practices of what we have identified as Stage 3 in the development of digital textual transmission. The wide availability of more user friendly application software (especially after the creation of XML out of the powerful but unwieldy SGML) has certainly lowered

---

[18] My forthcoming book 'Changing our textual minds: Towards a digital order of knowledge' (Manchester University Press), attempts to analyse some of these implications.

[19] The KNAW's Virtual Knowledge Studio was founded as part of a movement to counter such fears.

[20] This is not to suggest that the application of markup, which requires extreme discipline and rigidity, does not propagate new qualitative insights into the nature of the material being marked up.

the entry barriers. Equally, libraries and other institutions are now taking strides in offering 'lay' environments for the use of TEI and other markup for the transcription of a variety of primary sources.

Model B involves using the computer in ways that are primarily based on quantity. Now that we have become comfortable with text processing as a form of symbol manipulation, we can cast around for all sorts of new ways in which text may be subjected to computer treatment. Here the sciences can also offer inspiration, challenging us to turn their methods and inventions to unforeseen uses.[21] Often such ways of treating text will be based on statistical methods, requiring digitised materials in bulk (if only for training purposes). What sets this model apart from the others (at least from A and C) is that it offers a vista of yielding types of knowledge that are fundamentally different in the sense that they are not based on existing human knowledge or insights, or human analysis.[22]

Despite the fact that both Model A and Model B involve sophisticated use of computing power, they are conventional in the sense that they still involve mainly highly trained experts. The resulting knowledge is hierarchical, top-down. *Model C*, by contrast, involves 'using the user', and forms a significant departure from this conventional mould of expert knowledge creation. Apart from peer-to-peer horizontal collaboration, it entails vertical collaboration between experts and non-experts. It is of course hugely facilitated by the Web 2.0 developments of recent years. By analogy with the concept of 'Mode 2' production of knowledge[23] this can be called a 'Mode 3' type of knowledge production and research. Some interesting claims are beginning to be made for 'the wisdom of crowds' not least by James Surowiecki in his book by that name.[24] It will be a huge challenge to learn to trust such democratic forms of knowledge production.[25] That they are

---

[21] Roger Boyle gives a nice illustration of the possibilities in his paper elsewhere in this volume.

[22] In his contribution to this volume, John Lavagnino shows that a computer being fed quantitative statistical data may well come to a different verdict than a human taking into account (also) qualitative judgements of a kind that we will find it difficult to program a computer to take into account.

[23] Michael Gibbons et al., *The new production of knowledge*.

[24] James Surowiecki, *The wisdom of crowds*. Cf. also Cass R. Sunstein, *Infotopia*, and David Weinberger, *Everything is miscellaneous*.

[25] In her contribution to this volume Vika Zafrin cites Noah Wardrip-Fruin as saying that 'it turns out that the blog commentaries will have been through a social process that, in some ways, will probably make me trust them more'.

touted by many enthusiasts as being just as good if not better as the expert type might not necessarily help. However, the aspersions cast by scientists on long-familiar scientific types of knowledge production and research just might.[26]

Mode 3 type of knowledge production involves a radically different way of presenting knowledge. In fact it is a matter of publish data first, and select, edit, perfect them later.[27] In fact that may be identified as a general tendency: to publish information earlier, but defer its interpretation to turn it into knowledge. This is a social consequence of certain inherent properties of the medium, such as notably the fact that storage is cheap, copying costs nothing and doesn't exhaust the original, etcetera. In learning to value such more democratic forms of knowledge production we will have to make an even greater hermeneutic effort as we turn the data we consume into something that we feel can be called 'knowledge' in the scholarly sense.

This type of democratic knowledge creation by unnamed collectives might be termed 'amateur humanities'. What will be particularly interesting to see is if the knowledge resulting from this type of knowledge creation will be different from the knowledge derived from more traditional forms of knowledge creation, or whether the same 'truths' will turn out to be arrived at by different methods.

Lastly, *Model D* is a variation of Model C, which we might term 'using use'. It involves logging the use that is made of existing resources with a view to learning from it in ways that may improve those resources. Web users already frequently experience such automated analysis of queries and other forms of use; Amazon.com's cheerful greeting, followed by their equally cheerful and more or less apt buying suggestions, no doubt being among the most familiar encounters. Such logging should give us a better grasp of how our new knowledge instruments are being used, giving us a chance to improve them.[28]

Of course all of these models can—and do—occur in any combination. What all of them have in common is the need to do two things:

---

[26] Cf., for example, Sheila Jasanoff, 'Technologies of humility; Sal Restivo, 'Modern science as a social problem'; and Helga Nowotny's 'Transgressive competence'.

[27] See David Weinberger, *Everything is miscellaneous.*

[28] In fact, accountability systems and new research ethics increasingly demand such logging.

(1) We need to establish the 'true nature' of the computer (in inverted commas, because the computer's nature as a Universal Machine eliminates the possibility of inherent properties). This is the process I have termed *discovery*.

(2) We need to be very creative and proactive in helping develop the way we can use the computer for advanced text processing. This involves innovation, modelling, and enormous creativity. Only if we can articulate our demands can we hope to make any real progress. Acting on such well-articulated demands is what might be called *invention*.

I have drawn historical parallels because I believe that they contribute to a greater awareness of where we stand as humanities scholars in relation to the digital textual technology. The technology is not only new but likely to remain new for as long as we keep developing it—that is, forever. Yet culture demands continuity. Fortunately, the underlying structures are likely to remain stable; they will prevent irreparable breakdown of things like communication, knowledge generation, curation of our 'history', etc. I would suggest that the structure that is underlying knowledge transmission may well remain the Order of the Book for the foreseeable future. Even if this should just be a convenient metaphor that needs to be tested against an empirical practice, it would confirm once more the continuity of the critical role of humanities scholarship across any historical knowledge regime—including our own. If there is a lesson to be learned or a conclusion to be drawn, it would appear to be that we will need to keep reinventing our working methods—and thereby ourselves as well.

References

Bijker, Wiebe, *Bicycles, bakelites and bulbs. Towards a theory of sociotechnical change* (Cambridge, Mass. and London: MIT Press, 1995).

Blackmore, Susan, *The meme machine* (Oxford: OUP, 1999).

Darnton, Robert, 'What is the history of books', *Daedalus* (Summer 1982), 65–83.

Gibbons, Michael, Camille Limoges, Helga Nowotny, Simon Schwartzman, Peter Scott, and Martin Trow, *The new production of knowledge. The dynamics of science and research in contemporary societies* (London: Sage, 1994).

Havelock, Eric, *Preface to Plato* (Cambridge, Mass: Harvard University Press, 1963).

Jasanoff, Sheila, 'Technologies of humility. Citizen participation in governing science', *Minerva* 41, 3 (2003), 223–244.

Needham, Joseph (ed.), *Science and civilisation in China*, vol. 5: *Chemistry and chemical technology*, Part 1: Tsien Tsuen-Hsuin, *Paper and printing* (Cambridge: CUP, 1985).

Nowotny, Helga, 'Transgressive competence. The narrative of expertise', *European journal of social theory* 3, 1 (2000), 5–21.

Peters, John Durham, *Speaking into the air. A history of the idea of communication* (Chicago and London: University of Chicago Press, 1999).

Restivo, Sal, 'Modern science as a social problem', *Social problems* 35, 3 (1988), 206–225.

Saenger, Paul, *Space between words: The origins of silent reading* (Stanford: Stanford University Press, 1997).

Smith, Margaret M., 'Printed foliation. Forerunner to printed page-numbers?', *Gutenberg Jahrbuch* 63 (1988), 54–70.

Sperberg-McQueen, C.M., and L. Burnard (eds), *Guidelines for the encoding and interchange of machine-readable texts* (Chicago and Oxford: Association for Computers and the Humanities, 1990).

Sunstein, Cass R., *Infotopia. How many minds produce knowledge* (New York: OUP, 2006).

Surowiecki, James, *The wisdom of crowds* (New York: Doubleday, 2004).

TEI P5: *Guidelines for electronic text encoding and interchange*, eds Lou Burnard and Syd Bauman (Oxford, Providence, Charlottesville and Nancy: TEI Consortium, 2008).

Weel, A.H. van der, 'Changing our textual minds: Towards a digital order of knowledge' (forthcoming from Manchester University Press).

Weinberger, David, *Everything is miscellaneous. The power of the new digital disorder* (New York: Henry Holt, 2008).

PRESENCE BEYOND DIGITAL PHILOLOGY

Ernst D. Thoutenhoofd

A variety of digital developments are underway in text scholarship, but what wider implications are suggested by philology's aim to recover or reproduce (original) presence? What follows responds to Hans Ulrich Gumbrecht's philological concept of textual presence from a sociological perspective. It considers why a generative theory of presence matters to philological practice and beyond, submits philological means of producing presence to an analysis that contrasts Gumbrecht's typology of presence effects with Niklas Luhmann's influential theory of social systems, and places Gumbrecht's typology in the context of other digital scholarship studies that have also taken an interest in presence. The conclusion is that there is common cause in shared humanities and social science research on digital textuality and conceptualisations of presence.

Introduction

As we lay foundations for translating our inherited archive of cultural materials, including vast corpora of paper-based materials, into digital depositories and forms, we are called to a clarity of thought about textuality that most people, even most scholars, rarely undertake.[1]

Every technological repercussion and economic transformation threatens stratification by status and pushes the class situation to the foreground. Epochs and countries in which the naked class situation is of predominant significance are regularly the periods of technological and economic transformation.[2]

Wido van Peursen and I had planned some papers in the Text Scholarship and Digital Creativity Colloquium (in Amsterdam, October 2008) to reflect on Hans Ulrich Gumbrecht's proposals for changing philological theory and practice to take better account of material presence.[3] For my part, I had been intrigued by its dual concerns with power as

---

[1] J. McGann, 'Marking texts of many dimensions', 198.
[2] M. Weber, *Economy and society*, 983.
[3] H.U. Gumbrecht, *The powers of philology*, and *Production of presence*.

embodied instrumentalism and with materiality as a neglected aspect in an overly res cogitans humanities scholarship. Although materiality is also a key topic in the sociological sub-discipline of science and technology studies or STS, the 'presence' concept through which Gumbrecht set out to recover materiality is hardly known in social science,[4] except perhaps in relation to discussions about fieldwork methods. I happened on the particular casus of e-philology by way of meeting Wido van Peursen and Eep Talstra in person, a meeting that has led to a colloquium and this book. We share an interest in turning the increasing application of new technologies in humanities research into a new area of investigation for STS, since philology can probably benefit from the sensitivity to the social nature of research transformations that is characteristic of STS inspired work. STS scholarship reflects a rudimentary archeology of that sensitivity by way of Robert K. Merton's *The sociology of science* of 1973, Thomas Kuhn's *Structure of scientific revolutions* of 1962, and further back, Ludwig Fleck's *The genesis and development of a scientific fact*, originally published in 1935. STS studies now extend into critical analyses of the role of science in such matters as culture, policy and governance, while taking account of the processes and techniques that shape facts, values, technologies and understanding generally. The interaction of scientific and intellectual endeavour with politics and social organisation is a rich area of research, as is technological innovation. As part of its conclusions to date, STS inquiry has by and large insisted that scientific practices and outcomes are partly shaped by the material artefacts that it both uses and produces. STS scholars have more recently argued that it no longer makes sense to consider social agency a typically human attribute, and that the world (or, the various ways in which the world is conceived of) is at least partially determined by techniques and technologies acting, as it were, independently if not deterministically.[5] Given the colloquium's focus on new technologies, one set of papers were accordingly anticipated to expound on the processes of knowledge creation and representation in 'e-philology'.

---

[4] One notable exception is Niklas Luhmann's theory of social systems, of which more later.

[5] B. Latour, *We have never been modern*; K. Knorr Cetina, *Epistemic cultures*; R. Schroeder, *Rethinking science, technology and social change*.

Text, as a record of ideas, a means to construct author(ity), and material carrier of communication between humans has been central not only to philology, but to scholarship generally. Text is knowledge represented as matter: visible and revisitable, portable and measurable. As discipline focused on understanding texts and their transmission, philology is therefore a unique scholarly resource for understanding ways in which text alters under new sociotechnical conditions, but of course knowledge of text in philology is itself also a text, both epistemologically specific and formally encoded (theorised). Does new technology make philological approaches and insights into the nature of text more transparent for other scholars? What broader challenges, shared interests and opportunities emerge as text comparison becomes part of a wider move towards integrated forms of (collaborative) e-research and the multi-purposing of data collections? (From the colloquium invitation)

This invitation aimed at contributions able to externalise presumptions about methods and ways of doing that are normally buried inside the daily practice of e-philology. Combining Gumbrecht's focus on the production of presence with our own interest in the digital conditions of that daily practice seemed to offer us a way to test Jacques Derrida's claim that presence is usually 'dogmatically asserted [...] or set up as the absent goal of rational questioning' against digital environments.[6] As it turned out, absence of discussion about presence dominated the colloquium itself, as can be seen in the chapters of this book. Derrida's broader claim is of course that thought itself is constrained by reference to a rational form of reasoning that ties presence to time, contrasting past presence with future presence—and this claim too surely has a special resonance in the case of e-philology. As we will see by looking more closely at Gumbrecht's ideas, the time/subject incompatibility that Derrida refers to indicates a weak spot in Western metaphysics that characterises the production of presence in philological scholarship, since it is concerned precisely with always closing gaps that open between texts that persist through time and authors and readers who do not. An example of this is the longstanding practice of adding marginalia to texts that correct, critique, re-interpret or contextualise source matter.[7]

So, as Wido van Peursen has already noted in the introduction, we had explicitly referred to Gumbrecht's *Production of presence* in our colloquium invitation. I was therefore bemused to find that our explicit

---

[6] J. Phillips, *Deconstruction*, 194.
[7] See Ulrich Schmid's contribution elsewhere in this volume.

reference to those ideas inspired no shared reflection during the colloquium. This raised some further questions: does presence in digital texts (what we might call 'e-presence') matter to anyone? Is e-presence a topic worth thinking about? Even if it remained something barely analysed, is the unstated nature of presence a problem for scholarship? Is it indeed the case that the philological attributes of presence can be derived by reference to the traditional philosophical literature on the metaphysics of presence? In this chapter, an extension to a concluding colloquium talk that looked at images of biblical text that I had seen on my travels in Second Life (see Plates 26–29), I will maintain that current developments in e-philology and elsewhere are an invitation to look at such questions more closely in the social sciences.

Does a good theory of presence matter for e-philology?

A good theory of presence matters if we seek to better understand digital developments in scholarship. Jacques Derrida was not the first to tie the form that rational thought itself takes to presence in language (logos), although he may well have been instrumental in projecting an explicit symbiotic interaction between a particular conceptualisation of the historical passing of time, the endless reconstruction and re-interpretation of texts (which places philology at the heart of intellectual self-construction), and a distinctively Western metaphysics that strongly asserts individual subjects as 'authors'.[8] As Steve Clark has noted, Plato had much earlier contemplated, with sombre outcomes, the place of writing with respect to the transmission of ideas, denouncing it altogether for its failure to be tied to time, place and context.[9] Clark comments that ever since Plato, the opposition between speech and writing has been paradigmatic of the opposition between presence and absence that have co-structured Western metaphysics, along with other oppositions. Skipping centuries of philosophical musings—time, space and context are limited here—we can then return to Derrida's assessment. Formulating that assessment in sociological terms, writing associates with the exercise of power and control, not (only) explicitly but implicitly by writing being linked to political orders that marry onto-theological world-views (the promotion of texts in order to pre-

---

[8] J. Derrida, *Of grammatology*.
[9] S. Clark, 'Writing'.

serve a particular reading) with emancipatory potential. This suggests a tension of socio-political import, a means for controlling the masses and/or unleashing them, a tension that Ivan Illich and Barry Saunders described in socio-historical terms as 'the alphabetization of the popular mind' in a seminal text with that subtitle.[10] But as Clark concludes, Paul Ricoeur has more recently modified this onto-theology centred analysis by focussing on textual transmission as re-creative process. Transmission, in his view, involves translation work that permits world-making by accepting that things will always inevitably get lost anyway. In Ricoeur the absolute heterogeneity of textual interpretation becomes productive in a sense that Gumbrecht might acknowledge in spite of Ricoeur's lack of attention to material attributes. To my mind Eep Talstra (in this volume) comes closest to providing this missing material connection through his case study analysis, if we read his contribution as suggesting that corpus linguistic techniques—that after all have digital/material qualities—can offer some methodical counsel in relation to the translational heterogeneity discussed by Ricoeur.[11] However, the trouble with centering translations is that this then begs another question: why translate? Here Gumbrecht, as we will see, points to an innate hunger or desire that philologists have to 'recover by adding' (quite literally also in the sense of wanting to fill margins with notes). Clark has Ricoeur return to a model of the indeterminate role of translation in shaping the meaning of facts developed by Willard van Orman Quine,[12] who posited that besides two texts (original and commented or translated) a third ideal text must be assumed, a text that by nature of the task in hand can neither be attained nor ever

---

[10] I. Illich and B. Saunders, *ABC*.

[11] Further counsel is at hand in sociology, in particular with regards to Ricoeur's call for openness to the 'otherness' indicated by different readings. Willem Schinkel, for example, has pointed to the need to critically scrutinise even the most central concepts in an area of scholarship, focussing on the *casus* of social integration as topic in sociology itself, which presupposes the empirical and phenomenal validity of a 'society' in which one is expected to integrate. Instead, Schinkel contends that society is a rhetorical, discursive concept, so that we need a social theory *past society* if we are to understand the various forms of combatting otherness that lie behind the social integration impetus. See W. Schinkel, *Denken in een tijd van sociale hypochondrie*.

[12] For an historical contextualisation of Quine's ideas on language see Michael Losonsky's recent historical account in *Linguistic turns in modern philosophy*. Quine could see no sharp boundary between facts (truth) and meaning (logic), because according to Quine's view there can be no grounds by which to establish an inter-subjective standard for linguistic meaning, so that all translation (including that needed for translating facts into statements that matter to us) are necessarily indeterminate.

shared, yet invites pursuit for its superior value.[13] This is not the place to doubt these claims; but what I hope to show with these curt pointers is not just that philological activity contributes to ontological history— see Peter Øhstrøm and Ulrik Petersen's contribution in this book— but also that (its) conceptions of presence connect with historical and contemporary aspects of social shaping, and in particular with textual politics. This returns us to Hans Gumbrecht's dual preoccupation with the *powers* of philology and what we might now more precisely call *the means of producing* presence.

Gumbrecht's modern notion of meaning and pre-modern notion of presence

In the introduction to this volume, Wido van Peursen offered a humanities-based contextualisation of Gumbrecht's concept of presence, contrasting it with the views of Walter Benjamin and George Steiner in particular. I will instead focus on what Gumbrecht's concept of presence might teach us about digital text scholarship as a resource for social shaping. In *The powers of philology*, Gumbrecht posits 'a type of desire that, however it may manifest itself, will always exceed the explicit goals of the philological practices'.[14] The desire, it turns out, is for presence, for a physical connection to the things of the world, which on occasions that are not further specified may even be attained. Such instances then classify as 'presence effects'. The desire for presence is fed by an internal energy that the philologist is said to have. This may remind us, by the way, of Derrida's energy surplus, which is similarly indicated by a desire to translate.[15] Here is where Gumbrecht first mentions that philological practice is capable of releasing an oscillation between mind effects and presence effects, an oscillation that reminds him of contemporary definitions of aesthetic experience. This matrix of mind/meaning and materiality/presence then persists

---

[13] This assessment should be extended to include Sheila Jasanoff's call for technologies of humility, in this case text scholarship technologies that are able to reflect the indeterminate, transient character of all three of Quine's proposed versions of a text, since humble technologies—more than overbearing ones—would leave room to challenge theoretically supposed realities and therefore give better hopes of managing the practical consequences of the research directions that are taken. See Jasanoff, 'Technologies of humility'.

[14] H.U. Gumbrecht, *The powers of philology*, 6.

[15] S. Clark, 'Writing', 62.

throughout both of Gumbrecht's works. In sum, presence is a physical connection and a desire for it produces 'presence effects'—orientation to a discrete human being, a subject with individual agency, is presupposed by both aspects. In so far as philology can aspire to explicitly pursue this oscillation between meaning effects and presence effects (a programme that Gumbrecht advocates throughout *Powers of philology*), such a programmatic practice may give rise to a new intellectual style that is capable of challenging the very limits of the humanities, and presenting 'a beautiful and intellectually challenging fireworks display of *special effects*' (emphasis in original). While it is never made clear how presence might be put on empirical record, it is clear that Gumbrecht's conceptualisation is intellectually indebted to the philosophical phenomenology of Heidegger and Husserl. Phenomenology is historically interpreted by Gumbrecht as a late challenge to symbolic representation that would mark the birth of the Renaissance, a theory that is also commonplace in art history. Gumbrecht considers that Renaissance preoccupation with what later became formalised as semiotics replaced an earlier, medieval Christian culture centered on a collective, unquestioned belief in the possibility of God's real presence, as experienced during Mass; and it is precisely this earlier, phenomenal belief in presence that Gumbrecht seeks to recover. However, his historical claim gives Gumbrecht's notion of presence distinctly pre-modern credentials. Among the undifferentiating masses or *Gemeinschaft* of the pre-modern Christian era, a shared belief in the simple nature of things (along with their self-evident presence) could more readily be orchestrated than can now be recovered; in today's world some other type of production of presence is needed.

As Gumbrecht well realises, neither Western metaphysics, nor contemporary linguistics or social science lends much support for a pre-modern ontology of direct correspondence between subject and object. And indeed it also seems strangely out of kilter with the social systems theory posited by Niklas Luhmann that Gumbrecht notes in *Production of presence* was experienced by himself and kindred spirits in the 1980s as reflecting a sympathetic position, since it too (indeed) takes distance from philosophy's metaphysical focus on the subject. One problem, it seems to me, is that Luhmann does this for very different reasons. Since Luhmann posits that social systems are self-governing and disinterested systems of communications, his theory leaves no room for a subject at all, except in so far as subjects and agency are evidenced through the kinds of communications through which self-reference is constructed in social systems; a feature that is also known as pointing to primary or phenomenal social facts. Luhmann's radical constructivism, positing an originating 'difference' that separates a social system from something it henceforth conceives of as an environment and therein makes that environment un-knowably other, derives from Humberto Maturana and Francisco Varela's description of living organisms as autopoietic systems in neurobiology.[16]

But Luhmann's theory of social systems is a very long way off from recovering a pre-modern worldview for its more succesful construction of presence. In fact, Luhmann quite specifically notes that presence 'is the constitutive and boundary-forming principle of interaction systems, and presence means that people's being together there guides the selection of perceptions and marks out prospects of social relevance.'[17] I take this to mean that Luhmann sees presence as a principle of situated interactions, one that enables people to reach consensus about what matters and how that is relevant given the particular situation they see. If that assessment is reasonable, presence is akin to Pierre Bourdieu's collective obedience to rules, that lends a situation the structural coherence that is needed to make it work as a social event.[18] Given its reliance on subject agency, Gumbrecht's concept of presence would require some type of intersubjective agreement about 'what it is' that can then manifest itself through presence effects, but universal submission to cosmic forces seems an extravagant choice. A 'cosmic' concept of presence ties Gumbrecht to the longstanding metaphysical preoccupation that he criticises, namely the habit of distinguishing between the transcendental and the particular. Moreover, because of 'our' historical distance from a pre-modern mindset it leads him to a culturally charged interpretation. The outcome is that material presence becomes a historically backwards-interpreted symbolic meaning. At this point it will be instructive to recapture Gumbrecht's typology of presence and meaning interaction (which he indeed put forward as a cultural analysis) as a table, in order to make his particular cultural historical construction of meaning and presence visible as so many oscillations of a modern/pre-modern subject, or in effect as oscillations of the body politic.

---

[16] H. Maturana and F. Varela, 'Autopoiesis and cognition'.
[17] N. Luhmann, *Social systems*, 414–415.
[18] P. Bourdieu, *The logic of practice*, 53.

Table 1: Gumbrecht's typology of presence and meaning interaction
(pp. 80–85)

Category	Meaning effect	Presence effect
Self-reference	The mind (*res cogitans*)	The body (*res extensa*)
Culture	Subject-centered (eccentric)	Cosmological (inherent)
Knowledge	Interpreted (hermeneutics)	Substantive (revelation)
Communication	De Saussure's linguistic sign	Aristotle's substance/form coupling
Change	Instrumental transformation	Ecological rhythms
Order	Negotiated in time	Negotiated in space
Violence	Time converts it into power	Space converts bodies into obstacles
Progress	Welcome innovation	Unwelcome deviation from cosmology
Civic ritual	Parliamentary debate	Eucharist celebration

This typology does not enable a better understanding how these divisions are unified in the subject or what principles might guide inter-subjective agreement, and the book offers no empirical details about the production of presence in current philological practice; the discussion never departs from the abstract level of critical theory. However, what the typology does detail is alignment of presence effects with a social order conceived as onto-theocratic. Or to be more precise, the medieval Christian order that predated the Renaissance geo-politically, the Enlightenment intellectually, and the subsequent emergence of the experimental natural sciences scientifically. Gumbrecht must have anticipated this, making the Eucharist its prototypical civic ritual. Whatever Gumbrecht's concept of presence implies in terms of the socio-politics of humanist scholarship, in sociological terms it suggests a persistent disposition towards a nature of things that ignores the role of knowledge regimes and techniques in socio-political orders. While anachronistic, such a disposition might perhaps find expression in social movements (including religious ones),[19] but an association with underdog-positions then produces the dual paradox that this denies it the attribute of cosmic universality that it needs in order to

---

[19] Examples are Hamish, Jehova's Witnesses, and of course Creationism's literal interpretations of a central text.

be a manifest quality, and that a disposition that is persistent and cosmic nevertheless needs to be recovered and made meaningful through situated knowledge construction—Gumbrecht's presence is a cultural resource, not a cosmic à priori.

Absent from the typology of presence effects are attributes one might reasonably associate with materialities or even material qualities. In contrast with the general reliance on phenomenological discourse that runs through both *The powers of philology* and *Production of presence*, there is to my mind no clear articulation of the role of material qualities as substantive elements of agency, also not in the everyday excercise of philological work. What an STS student might be interested in could, for example, be an account of the tools, technologies and artefacts that a philologist would classify under work, and how the philological researcher would interact with those in a particular work-environment in order to 'do' (that is, construct) philology. The typology has therefore less to do with an empirically established account of human/object interactions—which Gumbrecht of course does not claim—than with an anachronistically conceived theoretical claim. In my interpretation, that claim draws on notable elements of two historically contrastive social orders, one scientific, the other pre-scientific, producing an oscillating 'effect' in philological agency that itself remains largely under-determined. So where Wido van Peursen, in the introduction to this volume, notes that Gumbrecht advocates real presence, the term 'real' has in Gumbrecht's case to be understood as cosmic, universalist, presupposed and beyond intellectual challenge, but not in any clear sense material or amenable to empirical verification. It has of course to be said that I may well offer a different critique than might be commonplace in text scholarship—the 'things' that matter to a discipline might get lost in cross-disciplinary translation. Gumbrecht himself returns to the question, 'What is presence?' at the close of *Production of presence*. The answer turns out to be 'as much out of reach as any other dream', whereafter Gumbrecht concludes that perhaps our media environment (technology) has the potential for 'bringing back some of the things of the world to us'.[20] Then follows, as if an afterthought, the second time that special effects are reported, this time not the end-product of prodigious philological desire but the product of engaging material technology: a compact disk

---

[20] H.U. Gumbrecht, *Production of presence*, 137–140.

and a screen. The material qualities of presence therefore remain an elusive attribute in Gumbrecht's discussion of presence, confounded even more by the digitality of the technologies referred to.

As some contributions to this book have shown, e-philological projects can focus on the material text with impressive outcomes, while other studies in e-philology concern digital enhancements for evidence-based emendation and criticism of the linguistic text. In e-philology these two orientations seem mutually shaping, even though at this time it is unclear how to formally interconnect the two streams of activity—or indeed separate them.[21] But while the distinction between material and virtual presence may be useful in discussing the transition to digital scholarship, it too skips an important prior question: is it any clearer what we mean by 'presence' when discussing digital text? Between materiality and digitality now lie new technologies of mediation that, in Gumbrecht's terms, have special effects on the construction of meaning and the experience of presence—on a phenomenological metaphysics. What Gumbrecht sought with *Production of presence* was a description of material presence as a distinctive element of understanding that humanities scholars risk losing sight of in their singular focus on interpretation, and in fact the same might apply to the application of computational possibilities that is unrestrained by theoretical reflection. If texts are treated as combinations of code, linguistic data, annotations and virtual reality (the digital approximation of material characteristics) then material presence—an effect which Gumbrecht has eloquently struggled to recover from both interpretation and constructivism—is radically transformed. However, from both STS and sociological perspectives it may be the case that the digital achievement of textual presence is best treated as generative interaction among various scholarship practices.[22] And there are already good examples of such a productive dialogue in the various conceptualisations of presence in research on tele-operative environments, web technologies for social interaction, and virtual ethnography. So, we will now turn to those and look for theoretical reflection on how presence can be operationalised in the development of digital products and techniques.

---

[21] See Ulrich Schmid's contribution to this volume.
[22] Pointing also to post-phenomenology in philosophy. See D. Ihde, *Postphenomenology*.

Neighbouring cases:
Non-philological scholarship and digital presence

At the time of writing, presence has become the research topic of an EU-funded consortium called PEACH or 'Presence Research in Action'.[23] PEACH concerns itself with developing a greater understanding of 'how we create the experience we call *reality*'. It focuses in particular on the experience of reality that is mediated through interaction technologies, and therefore the concern is with the cognition of humans and machines and the nature of their interaction. This research on presence takes as its starting point the experience of reality as a strictly cognitive event. A subject's sense of reality is taken to have a material base in our neurology, while the sense of reality that a computer constructs has a comparable material base in code. As a product of human and computer cognition, interaction between the two can be schematised on the basis of one important assumption, namely that 'in a sense, all reality is virtual';[24] this claim is then taken to support the contention that, in principle, computation can serve the same purpose as cognition, which is to mediate the experience of reality. Presence then becomes 'the experience of "being there" in a mediated environment'. Because the approach taken within PEACH abstracts the experience of reality to a core essence, consisting of physical events that adhere to the general patterns of lawful behaviour in a physical universe, it treats our experience of everyday reality as outcomes of neurological activity *per sé*. If we were to take digital presence as an experience that derives, in essence, from the combination of an ongoing flow of neural stimulation and computer code as our starting point for understanding the presence of text in e-philology, we would likely conclude that the experience of being in the presence of an ancient text (and perhaps its author) becomes ever more convincing as virtual environments become better at handling interaction and immersion. The contributions on the reconstruction of watermarks in paper, and the *InscriptiFact* and *Codex Sinaiticus* programmes of work could certainly be considered in that light.

Pavel Zahorik and Rick Jenison are psychologists at the University of Wisconsin who considered presence as part of a NASA-funded project

---

[23] http://starlab.info/peach/.
[24] W. IJsselsteijn, 'Elements of a multi-level theory of presence', 245.

on tele-operative and virtual environment developments.[25] Reviewing two ontological positions—one rationalist and one metaphysical—on the experience of reality, Zahorik and Jenison conclude that presence entails, first of all, an indivisible coupling between environment and action. Against the rationalist position, they argue that it treats situations as a combination of object attributes and actions that are subject to general rules of performance. The experience of situations might therefore be approved by fine-tuning the object attributes, so that actions appear more real. The problem that Zahorik and Jenison see is that such rational approaches establish 'a framework for problem solution in which one may apply a formal system of logic to arrive at a conclusion'. In other words, it is not clear how particularities and general rules are to be identified or described. A theory that is rational in Zahorik and Jenison's terms does not make clear how a Second Life avatar sitting cross-legged on a pillow relates to the general rules needed to make that action meaningful—and thereby 'real' as a form of meditation, for example. Zahorik and Jenison therefore object to conceptualisations of presence that are based on formal frameworks that act in accordance to coded rules. They do so on the basis that such frameworks fail to address the indivisibility of the physical and metaphysical levels of action, and that is a precondition for actions being meaningful. A rationalist position on the digital construction of presence can only assume that the existence of a 'real reality' is unimpeachable. It would therefore need to be supplemented with a kind of solipsism, since it is far from clear that there are ways of knowing other than through subjective and always mediated measures of presence.

Starting with a powerful combination of Heidegger's phenomenology and J.J. Gibson's ecological psychology,[26] Zahorik and Jenison instead argue for presence being tied to successfully supported action in an environment. As the psychologist Gibson himself put it, 'Things must be substantial before they can be significant or symbolic. A man must find a place to sit before he can sit down to think.'[27] This indivisible unity of environment and supported action has also been articulated by Heidegger—whose methods for examining the nature of presence

---

[25] P. Zahorik and R.L. Jenison, 'Presence as being-in-the-world'.
[26] J.J. Gibson, *The ecological approach to visual perception*.
[27] J.J. Gibson, *The perception of the visual world*, 199.

were informed by the interpretation of ancient texts—with reference
to the concept of 'readiness-to-hand'. As a reminder, while objects,
such as a bible, have no stable representation, whenever we wish to
read about Moses the object we call a bible is readily conceived, albeit
exclusively with reference to the action that it is meant to support.[28]
Indeed, the idea of successfully supported action in an environment
aligns (via Heidegger) with Gumbrecht's point about presence effects
having some kind of material agency. But interestingly, this agreement now straddles the distinction between material and digital text,
both conceived within a similar phenomenological discourse. Moreover, with cosmos replaced by environment, belief replaced by action,
and desire replaced by success, the construct now does approximate
Luhmann's conception of presence as a constitutive principle of interactive systems without apparent contradiction. Zahorik and Jenison
conclude that consequently, the presence of objects and self alike are
manifest exclusively as successfully supported action in an environment that 'reacts, in some fashion', to lawful action, so that the ecological coupling between perception and action must be considered
central to the notion of presence.[29] A holistic construction of presence
in text scholarship would therefore likewise call for a strong feedback
mechanism that tightly couples e-philological precepts and digital
text environment. Yet that in itself is not enough, since presence
should be a collective and not a private achievement if it is to matter
in scholarship.

In her PhD thesis, new media scholar Caroline Nevejan undertook
to describe an ecological psychology in computer-mediated communication by positing 'the thinking actor', who experiences multiple
presences within a four-dimensional space that includes the actor, the
time and the location as the first three dimensions and the range of
possible actions that are supported by that space as the dynamic and
unpredictable fourth dimension.[30] Within that schema, the natural
presence of the actor (what in virtual gaming would be called the first-
or real-world actor) is the factor of distinction, the form of presence in
which 'catharsis takes place.'[31] As is the case in 'natural presence', the

---

[28] M. Heidegger, *Being and time*, 98.
[29] P. Zahorik and R.L. Jenison, 'Presence as being in the world', 87.
[30] C. Nevejan, *Presence and the design of trust.*
[31] Taken from the online version of Nevejan, *Presence and the design of trust*, see
www.being-here.net/article-886-en.html.

psychological state of the actor influences the experience and hence the construction of presence. Nevejan supplements what Zahorik and Jenison omitted, namely the crucial fact that an environment rarely contains just one actor, hence 'for the accomplishment of an act, an actor is dependent on the work of other actors, each of whom are psychic beings operating according to dynamic rules. Under those conditions, incommensurability between the practices of agents striving after individual objectives have to be presupposed', so that successfully supported action needs to account for communities of practice that have developed intricate mechanisms for negotiating incommensurability, which might be described as the unlikelihood of actors sharing the same experience of reality or being 'truly co-present.' In sum, it is not necessarily always clear how success, as an outcome of action, is to be defined, experienced or acknowledged. The various constraints on action that make actors' noses point in the same direction point to social sources of trust. Emulating the collective orientations that might incite actors to place their trust in the world that is being constructed will present very particular challenges for computer-coded environments. Nevejan's work therefore achieves a necessary shift from an ecological psychology to an ecological *sociology* of mediated presence. Here language itself can take centre-stage again, since taxonomies— including lexicons and conceptual schemes—are the necessary consequence of the ways in which actors interact. Taxonomies can perhaps best be considered the historical database of negotiated incommensurability among types of mediated action. Such taxonomies offer a systemic, rule-governed (in the sociological sense noted earlier) means for achieving consensus and collaboration among actors who cannot otherwise have presence at all.[32] If we metaphorically conceived of textual criticism as theatre, presence is the result of those practical enactments that have needed that theatre as a space in which to act, therein accrediting the theatre's particularities.

Presence and technology is also a topic of investigation in virtual ethnography. Research by Anne Beaulieu in particular includes attention to how ethnographic study of mediated interaction (such as internet communications) might be made meaningful.[33] Ethnographic

---

[32] As Luhmann notes, 'the greater the technical influences on situations [...] the more compelling, but also the more autonomous!, the determination of social relevance', 415.

[33] A. Beaulieu, 'Mediating ethnography'.

accounts are, as the ethnographer James Clifford noted, 'always caught up in the invention' of cultures.[34] One of the challenges that ethnographers however face when studying the internet is that it seems already inscribed. As Beaulieu notes, internet interaction is predominantly— although perhaps increasingly less so—seen as textual. The internet therefore appears to ethnographers much the same as classical texts do to philologists: as literary output in which only traces of the social interaction that went into their making have remained. Hence ethnographers and philologists seem to pose similar questions: about meaning and presence in texts that are already there, and about how these texts interconnect with texts that result from studying them; this also includes similar concerns about issues of authenticity, validity and of course how to conceive of presence itself. The conclusions that pertain with respect to the notion of presence are equally less certain here, but not therefore any less pertinent. In particular, Beaulieu points to the central role of strategies of objectification that researchers studying the internet will use. In strategies of objectification the researcher uses technological mediation to remove themselves from the research site, which has the effect of turning the research target into a self-referencing object. Objectification is a notable attribute of scientific method, with scientific publications mostly limiting the use of personal pronouns and narrative voice to similar effect. Likewise, using internet technologies ethnographers may, for example, take on the role of 'lurker', performing a semi-presence that involves attending to but never participating in opportunities for exchange in chat-rooms or online games.

The comparison between these various conceptions of presence with the re-configuration of ancient texts into corpora and categorising them with reference to the technologies used (including visual representation, linguistic tagging, and emendation) might all class as examples of objectification strategies. In conclusion, whatever form presence takes is at one and the same time an artefact of scholars' construction of the object of research and an attribute of the kinds of social interaction that their research strategies can make visible.[35] But

---

[34] J. Clifford, 'Introduction', 2.
[35] VKS colleagues have more recently proposed the concept of focale (defined as sustained attention to a research topic) as replacement in e-research for the traditional notion of being 'in the field' (A. Beaulieu et al., 'Not another case study', 682).

however innovative, forward-looking or new, research strategies and research aims are governed by knowledge traditions that have long both divided and ruled the academy. They furnish durable, change-resistant textual resources for maintaining status quo across many areas of social differentiation. So that while new research strategies re-combine with emergent text technologies and offer novel opportunities for conjoining 'how we understand' with 'how we construct' presence in the world, these new opportunities are subject to social constraints that are much harder and slower to change, and that also find expression in the techniques and technologies that are being used for researching presence.

Summary

My own conclusions drawn from all of the above follow from the projection that philology is a (re-) inventive scholarly engagement with a metaphysical incompatibility between time and subject, pursued in a manner that is organised by insistent reference to the ideal-type of an original or 'source' text that must have commenced the textual transmission of ancient ideas. The intellectual transformation of this longstanding analytical attention to the textual origins of ideas under digital conditions of production makes 'e-philology' a highly relevant casus for social science analysis, as the latter is increasingly concerned with the heterogeneous development of ideas streaming among actors (humans and machines), and distributing in space and time by way of digital techniques and technologies. It now seems clear that advanced digital techniques, such as used in corpus linguistics and e-philology, offer new approaches for capturing the heterogeneous nature of knowledge-translations across dimensions of place, time and purpose. The variable and often presupposed concept of presence, and the ways in which presence is (im-) materialised as much as operationalised in various forms of digital scholarship, I think foreshadows the processes through which the metaphysically closely related concept of knowledge will be rethought in digital scholarship more generally.

Conceptions of presence connect with aspects of social shaping, and most directly in textual politics. This is most aptly captured to my mind in Gumbrecht's references to the *production* of presence. I have submitted that particular notion of production to a critique of its ahistoric credentials and have instead proposed a conjoining of humanities and social science forces, in the form of wider attention to *the means* of producing presence, in my own focus on e-philological tools and techniques as a case in point. I have further noted that computational possibilities and futures that are unrestrained by theoretical reflection risk losing sight of material presence. And similarly, where text is treated as an exclusively digital matter (such as text created as part of web-02 activities) then material presence is radically transformed. Since we can at least be certain that there will be more and less powerful, effective and humble means to produce presence, the risk is in leaving the future means of producing past presence to opportunistic analyses in hindsight. For all the above reasons, I have suggested that the digital achievement of textual presence is best treated as generative interaction among various forms of text creation and circulation. And finally I have noted some potentially productive similarities in patterns of thinking and analysis across disciplines involved in researching presence, pointing en route to the very similar conception of the internet among philologists and virtual ethnographers, both fields of inquiry seeing in the internet an overwhelming textual output in which only traces of the original producing relationships and activities that go into its constitution have remained.

Concluding remarks

The considerations presented here suggest that there is reason to investigate the interaction among scholars coming to terms with their digital scholarship (including new objects and research strategies), and pay close attention to the digital-era conceptualisations of presence, or 'e-presence'. But there needs also to be good common *cause*, a clear social objective, for presence to be visible to scholarship. Academic expertise still qualifies (and tends to present itself) as a stable power elite, and therefore scholarship presents a much more general problem with regards to its mission to understand and innovate the world.[36] With respect to the potential for sociological inquiry, there is a wide-open terrain in analysing how e-philology can help drive social change in the wake of digital transitions that span the social system and its vast range of textual communications. Conversely, e-philology and its

---

[36] S. Restivo, 'Modern science as a social problem'.

products are already participating in a 'third-wave' of mature science in society.[37] Its expertise and findings compete with expertise-claims and popular counter-technologies (such as Wikipedia) that have surfaced outside academia among hobbyists, interest groups, social movements and non-research institutions. How to defend, for example, the robustness of e-philological claims, or the stable qualities of research objects that are being constructed, or restrictions on online access to the objects that are put online are likely to become matters of more public concern in due course, transcending disciplinary interests and inseparable from questions about ethics, accountability, re-usability, availability, curation and participation—in sum, about public interest in the politics of text and textual presence.

Literary critic and text scholar Jerome McGann's comment that great clarity of thought is needed in laying the foundations for digital text repositories readily fits the themes addressed in this book. I have deliberately contrasted it at the start of my assessment of presence in e-philology with a much older quote that derives from an essay in classical sociology. The claim about a strong correlation between technological innovation and class upheaval made by German sociologist Max Weber (1864–1920) was posthumously formulated by his students, but it was an important principle in his assessment of the economy's role in social organisation.[38] Weber claimed that stable arrangements of stratification become most acutely challenged and disturbed when there is significant repercussion from technological innovation—such as currently the digital transition of texts. As book and literature scholar Adriaan van der Weel has argued in this same volume, we can at this time not expect to forecast what wider changes accrue with respect to text technologies and products, but there are well-analysed precursors to textual innovations that while minor in themselves had a considerable social effect. Examples are the gathering of sheets in codex form, the habit of adding word-space in early scriptoria, and the evolution of page numbering in printed texts. All these innovations, as Van der Weel notes, have helped to shape books into the 'reading machines' that have made mass-literacy and

---

[37] H.M. Collins, and R. Evans, 'The third wave of science studies'.

[38] Sociologists are nowadays less likely to foreground class distinctions because that concept was originally tied to purely economic activity leading to social stratification, a belief that is no longer widely held. What has persisted however is sociological interest in social differentiation and how it develops over time.

emancipation possible, along with supporting functions such as referencing and indexing. Collectively, these various textual innovations are thought to account for the most pivotal conventions of communication that apply to the present day—indeed, they can be said to define a bookish order of things. These defining consequences of earlier textual innovations foreshadow the impact that digital text technologies might eventually have on the future ordering of things *and* ideas.[39] At the same time, research technologies and scholarly expertise are themselves privileged forms of agency, so that we can expect to find motivated connections between for example e-philology as a network of professional expertise, constructions of digital text culture, and claims about presence. Because of the range in topics and expertise that it contains, this book offers many excellent starting points for an analysis of such motivated connections between future order and current action. It also suggests that a coupling of reflexive theoretical imagination and text scholarship/criticism is needed for analysing the effects of the philological search for presence on meaning and practice.

References

Beaulieu, A., 'Mediating ethnography. Objectivity and the making of ethnographies of the internet', *Social epistemology* 18, 2–3 (2004), 139–163.

——, A. Scharnhorst, and P. Wouters, 'Not another case study. A middle-range interrogation of ethnographic case studies in the exploration of e-science', *Science, technology and human values* 32, 6 (2007), 672–692.

Bourdieu, P., *The logic of practice* (Cambridge: Polity Press, 1990).

Clark, S., 'Writing', *Theory, culture and society* 23, 2–3 (2006), 60–63.

Clifford, J., 'Introduction', in *Writing culture. The poetics and politics of ethnography*, eds J. Clifford and G.E. Marcus (Los Angeles: University of California Press, 1986).

Collins, H.M., and R. Evans, 'The third wave of science studies. Studies of expertise and experience, in *Social studies of science* 32, 2 (2002), 235–296.

Derrida, J., *Of grammatology* (Baltimore: Johns Hopkins University Press, 1974).

Fleck, L., *The genesis and development of a scientific fact* (Chicago: Chicago University Press, 1979).

Gibson, J.J., *The ecological approach to visual perception* (Hillsdale, New Jersey: Lawrence Erlbaum Associates, 1979).

——, *The perception of the visual world* (Cambridge, Mass: Riverside Press, 1950).

Gumbrecht, H.U., *The powers of philology. Dynamics of textual scholarship* (Urbana, Illinois: University of Illinois Press, 2003).

---

[39] I use the term 'order' to refer to any durable arrangement of social differentiation within a given setting.

——, *Production of presence. What meaning cannot convey* (Stanford: Stanford University Press, 2004).

Heidegger, M., *Being and time* (London: Harper Collins, 1962).

Ihde, D., *Postphenomenology. Essays in the postmodern context* (Evanston, Illinois: Northwestern University Press, 1993).

IJsselsteijn, W., 'Elements of a multi-level theory of presence. Phenomenology, mental processing and neural correlates', in *Proceedings of PRESENCE 2002.* (Porto: Universidade Fernando Passoa, 2002), 245–259.

Illich, I., and B. Sanders, *ABC. The alphabetization of the popular mind* (London: Marion Boyars, 1988).

Jasanoff, S., 'Technologies of humility. Citizen participation in governing science', *Minerva* 41, 3 (2003), 223–244.

Knorr Cetina, K., *Epistemic cultures. How the sciences make knowledge* (Cambridge: Cambridge University Press, 1999).

Latour, B., *We have never been modern*, tra. C. Porter (Cambridge, Mass: Harvard University Press, 1993).

Losonsky, M., *Linguistic turns in modern philosophy* (New York: Cambridge University Press, 2006).

Luhmann, N., *Social systems* (Stanford, California: Stanford University Press, 1995).

Maturana, H. and F. Varela, 'Autopoiesis and cognition: The realization of the living', in *Boston studies in the philosophy of science nº42*, eds R.S. Cohen and M.W. Wartofsky (Dordrecht: Reidel, 1973).

McGann, J., 'Marking texts of many dimensions', in *A companion to digital humanities*, eds S. Schreibman, R. Siemens, and J. Unsworth (Oxford: Blackwell, 2004).

Nevejan, C., *Presence and the design of trust* (PhD thesis; Amsterdam: University of Amsterdam Press, 2007), also at www.being-there.net.

Phillips, J., 'Deconstruction', *Theory, culture and society* 23, 2–3 (2006), 194–195.

Restivo, S., 'Modern science as a social problem', *Social problems* 35, 3 (1988), 206–225.

Schinkel, W., *Denken in een tijd van sociale hypochondrie. Aanzet tot een theorie voorbij de maatschappij* (Kampen: Klement, 2007).

Schroeder, R., *Rethinking science, technology and social change* (Stanford, California: Stanford University Press, 2007).

Weber, M., *Economy and society. An outline of interpretative sociology* (Berkeley: University of California Press, 1978).

Zahorik, P., and R.L. Jenison, 'Presence as being-in-the-world', *Presence. Teleoperators and virtual environments* 7, 1 (1998), 78–89.

AUTHOR INDEX

Alkhoven, Patricia 34
Ambrose xiv
Andersen, Claus Asbjørn 58, 61, 70
Aquinas, St Thomas 18, 100, 262
Aristotle 66
Augustine xiv

Babbage, Charles 257
Baron, N. 229
Bazerman, Charles 81
Beaulieu, Anne 283–284
Bebb, Bruce 107
Benjamin, Walter 6, 9
Berners Lee, Tim 258
Blake, Norman 102
Boccaccio, Giovanni 211
Borges, Jorge Luis 102, 104
Bourdieu, Pierre 276
Bowers, Fredson 101
Boyle, Roger 4, 8, 17, 24, 25, 265
Breay, Claire 180
Brockett, A. 142
Brown, Dan 23
Brown, Tim 180
Burman, L. 85
Busa, Roberto 18, 100, 262

Caesar, Julius 22
Carotta, Francesco 22
Cervantes 103
Clark, Steve 272
Clifford, James 284
Cohen, Daniel J. 108–109
Crystal, David 12, 17, 19, 21, 25

Dahlström, Mats 6, 9, 10–11, 12, 13,
  19, 22, 24
Dante 178, 210
Darnton, Robert 259
Darwin, Charles 178
Dearing, V.A. 116
Dees, A. 116
de Hamel, Christopher 189
de la Mare, Walter 106, 107, 109
Derrida, Jacques 271, 272, 274
de Saussure, Ferdinand 249
Descartes, René 5
Donne, John 255

Doorn, Peter 34
Douglass, Jeremy 218
Duguid, P. 89

Eccles, Mark 109
Edwards, Mark 218
Eggert, Paul 211
Ellis, Willem 119

Fleck, Ludwig 270
Friedrich, Caspar David 7
Frohmann, Bernd 85

Gants, David L. 134, 144
Garces, Juan 180
Gesenius, W. 50
Gibson, J.J. 281
Göckel, Rudolf 61
Goethe, Johann Wolfgang von 103
Goetz, Nicolas 255
Greetham, D. 81
Greg, W.W. 116
Gruber, T.R. 70
Gumbrecht, Hans Ulrich 4–6, 269–290
Gutenberg, Johannes 190, 254–255, 259

Händel, Georg 7
Hansell, E.H. 177
Harmsen, J.H. 54
Havelock, Eric 256
Hawthorne, N. 101, 105, 106, 108
Heidegger, Martin 275, 281
Hennig, Willy 119
Herbert, George xii–xiii
Hiary, Hazem 4, 8, 17, 24, 25, 136, 145
Homer 103
Hooykaas, R. 63
Horace 256
Houghton, Hugh 180
Housman, A.E. 106, 107
Huang, Y. 132
Hunt, Leta 9, 11, 19, 21, 24
Husserl, Edmund 275

Illich, Ivan 273

Jasanoff, Sheila 274
Jenison, Rick 280–282, 283

Jerome xiv
Jongkind, Dirk 177, 184
Joüon, P. 50

Kane, George 104
Kanzog, K. 84
Kautzsch, E. 50
Kepler, Johannes 60
Kevern, Rachel 180
Kircz, Joost 12
Kjellman, U. 94
Kuhn, Thomas F. 270

Lachmann, Karl 113, 115
Lake, Kirsopp 177
Lake, Silva 177
Lass, Roger 102, 103, 106
Lavagnino, John 13–14, 18, 19, 24, 265
Leavis, F.R. 8
Locke, John 225
Lombard, Peter xiv
Lorhard, Jacob 16, 57–76
Losonsky, Michael 273
Lovelace, Ada 257
Luhmann, Niklas 269, 270, 275–276
Lundberg, Marilyn 9, 11, 19, 21, 25

Maas, Paul 115
Master of the View of Saint Gudule xii
Maturana, Humberto 276
McCarthy, John 70
McGann, Jerome 102, 287
McKendrick, Scot 180
Melville, Herman 106
Merton, Robert K. 270
Middleton, Thomas 100, 109
Milne, H.J.M. 177
Muraoka, T. 50
Myshrall, Amy 177, 180, 186

Nakra, Teresa 4, 131
Needham, J. 63
Nevejan, Caroline 282–283
Newton, Michael 223
Ng, Kia 145

Øhrstrøm, Peter 16, 18, 20, 21, 24, 58, 61, 63, 70, 274

Parker, David 10, 12, 22, 25
Parker, Hershel 107
Pavelka, K. 136
Pico della Mirandola, Giovanni 211
Pizer, Donald 101

Plato 256, 272
Poli, Roberto 70
Popper, Karl 114, 122
Pordenone, Andrea Galvani 138

Qimron, Elisha 149
Quentin, Dom H. 116

Ramus, Peter 63–64
Ricoeur, Paul 273
Riva, Massimo 211
Robinson, Peter 83, 178
Rolewinck, Werner 255
Rosenzweig, Roy 108–109

Salemans, Ben 13, 14, 15, 24, 99
Sandborg-Petersen, Ulrik 16, 18, 20, 21, 24, 274
Sappho 256
Saunders, Barry 273
Schärfe, H. 58, 61, 70
Schinkel, Willem 273
Schmid, Ulrich 21, 22, 25
Scrivener, F.H.A. 177
Shakespeare, William 103, 105, 107, 250
Shanks, Hershel 149
Skeat, T.C. 177
Sontag, Susan 5
Sowa, John 71
Spencer, Matthew 100
Steding, S.A. 84
Steiner, George 6, 8
Surowiecki, James 265

Talstra, Eep 2, 6, 9, 11–12, 13, 14, 15, 18, 23, 270, 273
Tanselle, G. Thomas 102
Taylor, Gary 100
Terras, M. 132
Thoutenhoofd, Ernst 1, 11, 17, 25
Timpler, Clemens 61–62, 64
Tischendorf, Constantin 175

Uckelman, Sara L. 71, 72
Uspenski, Porphyry 175

van der Weel, Adriaan 14, 15, 18, 22, 23, 25, 287
van Ghemen, Govert 117, 122
van Orman Quine, Willard 273, 274
van Peursen, Wido 269, 270, 271, 274, 278
Varela, Francisco 276

Villani, Giovanni   211
von Kleist, Heinrich   84

Wardrip-Fruin, Noah   217–218, 265
Wattel, Evert   118
Weber, Max   287
West, Martin L.   191

Zafrin, Vika   17, 20, 21, 25, 265
Zahorik, Pavel   280–282, 283
Zarri, G.P.   116
Zimmerli, W.   52
Zuckerman, Bruce   9, 11, 19, 21, 25

SUBJECT INDEX

acceleration (of scholarly practices by
the computer) 14, 16, 32–34, 262
amateur humanities 264–267
annotation 20, 209–211, 220–221
anonymity 250–251
artificial intelligence 131–134
authority 21–23
author-reader relationship 256,
261–262
authorship 248–250
  – multiple xiv–xv, 39–40, 181–195,
  249

back-lighting 136, 138
book of the heart xi–xii
bookwheel 16–17

cladistics 119–125
codex ix–xvi, 167, 174, 189–190,
200–201
Codex Sinaiticus 173–187
codicology 6, 129–144
collaboration (in the production of
ancient texts) 54; see also authorship,
multiple
  – online 16, 207–226
communication, mediums of 229–251;
see also text as communication vehicle
and digital communication
content vs form x–xi; see also meaning
vs presence
continuity, see functional continuity
(from one medium to another)
corpus methods 13–14, 99–110, 230,
273; see also corpus linguistics and
digital corpus
corpus linguistics 230, 273; see also
text as source for study of language
creativity 24, 267; see also digital
creativity

deductive vs inductive 113–125
digital
  – vs analogue 3, 190, 253, 257, 262
  – communication 3–4, 229–251,
  253–254, 257–260
  – corpus 10–12, 243

– creativity 2–4
– edition, see edition, digital
– images 11–12, 81, 83, 88, 136–144,
149–170, 176–180, 199–201
– object 9–12, 196–197, 214
– transcription 177–187, 197–203
digitisation
  – of libraries 87–96
  – critical 90–94
  – mass 88–90
discovery, see invention vs discovery
distributed networks 207–226
document, see text vs document
document structure
  – of ancient texts 35–38, 41–42
  – of digital texts 212

Easter Wings (Herbert) xii–xiii
editing, see text editing
edition
  – digital 83–96, 178–180, 196–202
  – printed 18–19, 84
  – as interpretation, see interpretation,
  scholarly edition as
emendation 99–110
emoticons 238–239
epigraphy 149–170

form, see content vs form
functional continuity (from one medium
to another) 14–15, 254–258, 261–262,
267

Google Books 11, 88–90, 106–107

hermeneutics, see interpretation
hypertext 16, 57–74, 236, 243, 246, 258
HTML 73, 258; see also hypertext

idiographic vs nomothetic 82–83
image
  – digitisation, see digital images
  – editing 11–12, 81, 91
  – documentation and distribution
  151, 157–169, 178–180
  – as interpretation, see interpretation,
  images as

imaging technologies 136–142, 151–157
image-oriented vs text-oriented editing 83, 88
imitation
 – of technological functionality of one medium in another 14–15, 23, 31–34, 253, 262
 – of 'real' objects in Second Life 11
 – of writing or speech in digital communication 232
inductive, see deductive vs inductive
interpretation 4–7, 266, 272–279
 – scholarly edition as 79–86
 – tradition of 37, 43–51
 – transcription as 185–187
 – images as 7, 11–12
 – medial bias as 261
invention vs discovery 254–258, 267

knowledge
 – definition of 20, 66
 – instrument, the computer as 264–267
 – creation of 20–21, 25
 – hierarchical vs democratic creation of 22, 262–263, 265–266
 – Mode 2 of knowledge production 265–266
 – Mode 3 of knowledge production 265–266
 – representation 3, 16, 20–22, 70

libraries, see digitisation of libraries
Lachmann, method of 113–125
Lanseloet van Denemerken 117–125
logic, history of 57–74
Lombard, Peter xvi
Lorhard, Jacob 57–74

Mahdiyya Quran 138–142
meaning 261, 273
 – vs presence x–xi, 1–26, 274–279
metadata 88–01
model-based approach 137–139, 144

nomothetic, see idiographic vs nomothetic

objective vs subjective 12–13, 80, 82, 113–125
ontology 57–74
open source 225–226
Order of the Book 23, 253–268

paper manufacture 134–144, 182

persistence (of written language) 240–241, 256, 271
palimpsest 162–163
pragmatics (linguistic) 243–245, 248
presence 1–26, 279–288; see also meaning vs presence

reading machine (the book as) 255, 259, 287
Reflection Transformation Imaging 155–156, 167

scholarly vs scientific 12–14, 79–96 et passim
Science and Technology Studies 254, 270, 278–279
social agreement 276–277
social order 277–278
social tagging 210, 220–221
stemma-building 15, 99, 113–125

TEI (Text Encoding Initiative) 263
text
 – as artefact 6, 9, 149–170
 – as communication vehicle 2, 20, 35, 271
 – comparison 2–4, 31, 33, 232
 – as literary composition 35, 37
 – concept of 6, 12, 17, 19, 35, 164, 209, 229–251
 – vs document 6, 34, 38–39, 83
 – encoding 12, 40–44, 84, 88–89, 211–216, 263; see also HTML, XML, TEI
 – genealogy, see stemma-building
 – vs images, see image-oriented vs text-oriented editing
 – politics 274, 287
 – as source for study of language 35, 37–38; see also corpus linguistics
 – syntactic structure 40–42, 232
 – transmission 35, 99–110
textual criticism 32, 80, 95, 113–125
 – history of 3, 53–54
text editing 79–86, 94–96, 99–110; see also edition
 – analogue 190–196; see also digital vs analogue
 – Alexandrian vs Pergamanian 81, 93
 – collaborative 207–226
 – critical, see transcription vs critical editing
tradition, see interpretation, tradition of

transcription, *see* digital transcription *and* interpretation, transcription as
 – vs critical editing   102
 – errors   102, 109
transformation
 – of concepts and categories   19–20
 – in scholarly practices   83–86, 201
 – medial   261–262
transparency, medial   260–261
turn-taking   237–238

Universal Machine   258–259, 267

virtual ethnography   279–280, 283–284

watermarks   134–144
Web 2.0   263–265
Wikipedia   22, 249, 287
world view   57–74

XML   72–73, 212

COLOR ILLUSTRATIONS

Plate 1: The dead Christ lamented by Mary, St. John and Joseph of Arimathaea (?),
Lamentation from a Book of Hours, Southern Netherlands ca. 1450. The Hague,
Koninklijke Bibliotheek (76 G 22, fol. 19v).

Plate 2: The last judgment. Fresco. Albi Cathedral, ca. 1490.

Plate 3: Young man holding a book. Master of the View of Sainte-Gudule. Oil on wood. Netherlandish, ca. 1485. New York, Metropolitan Museum of Art (acc.no. 50.145.27).

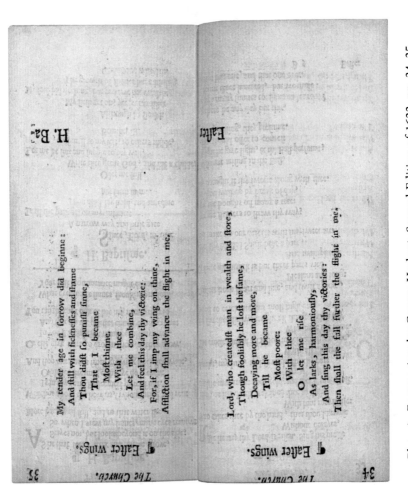

Plate 4: *Easter wings* by George Herbert. Second Edition of 1633, pp. 34–35.

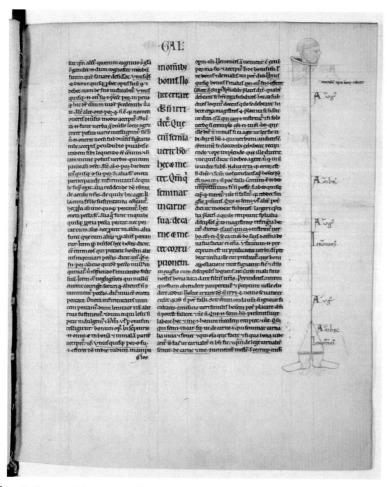

Plate 5: Page originating in the scriptorium of twelfth-century cleric Peter Lombard, Bishop of Paris (Paris, Bibliotheque Nationale, ms.lat.14267, fol. 26r).

עֶזְרֵנוּ בְּשֵׁם יְהוָה עֹשֵׂה
שָׁמַיִם וָאָרֶץ:
שִׁיר הַמַּעֲלוֹת: 125
הַבֹּטְחִים בַּיהוָה כְּהַר
צִיּוֹן לֹא יִמּוֹט לְעוֹלָם
יֵשֵׁב:
יְרוּשָׁלִַם הָרִים סָבִיב
לָהּ וַיהוָה סָבִיב לְעַמּוֹ
מֵעַתָּה וְעַד עוֹלָם:
כִּי לֹא יָנוּחַ שֵׁבֶט הָרֶשַׁע
עַל גּוֹרַל הַצַּדִּיקִים לְמַעַן
לֹא יִשְׁלְחוּ הַצַּדִּיקִים
בְּעַוְלָתָה יְדֵיהֶם:
הֵיטִיבָה יְהוָה לַטּוֹבִים
וְלִישָׁרִים בְּלִבּוֹתָם:
וְהַמַּטִּים עֲקַלְקַלּוֹתָם
יוֹלִיכֵם יְהוָה אֶת פֹּעֲלֵי
הָאָוֶן שָׁלוֹם עַל
יִשְׂרָאֵל: שִׁיר 126
הַמַּעֲלוֹת בְּשׁוּב יְהוָה
שִׁיבַת צִיּוֹן הָיִינוּ
כְּחֹלְמִים:
אָז יִמָּלֵא שְׂחוֹק פִּינוּ
וּלְשׁוֹנֵנוּ רִנָּה:
אָז יֹאמְרוּ בַגּוֹיִם
הִגְדִּיל יְהוָה לַעֲשׂוֹת

a uxiliu nrm innoie domini.
qui fecit celum oe terram.
CANTIC GRADVV.
Qui cofidunt in dno sic mon-
sion. non mouent in
eternum habitabit.
Iherlm montes incircuitu eius.
oe dns incircuitu ppli fui
modo -z usq; ineternum.
Quia non reqescet uirga ipietatis
sup forte ustoru. -z non
mittant uus tu ma
iniqtate manus suas.
Benefac domine bonis
-z rectis corde.
Qui aut declinant ad puitate
suas. deducet eos dns cu his
q opantu iniqtate. ax sup
drael. CANTICVM
Iu suertere dns GRAD
captiuitate fion. facti fum
quasi sommantes.
Tunc implebit risu os nrm.
oe lingua nra laude.
Tunc dicent ingentibus
magnificauit dns face cunctis.

Plate 6: Polyglot Psalter, fol. 266. Manuscript on parchment. Northern Europe, 12th century. Leiden University Library (UBLWHS.BPG.49A).

Plate 7: Bookwheel, Bibliotheca Thysiana, Leiden.

Plate 8: *Ogdoas Scholastica, continens Diagraphen Typicam artium: Grammatices (Latinae, Graecae), Logices, Rhetorices, Astronomices, Ethices, Physices, Metaphysices, seu Ontologiae* by Jacob Lorhardt. Sangalli: Straub 1606, frontispiece.

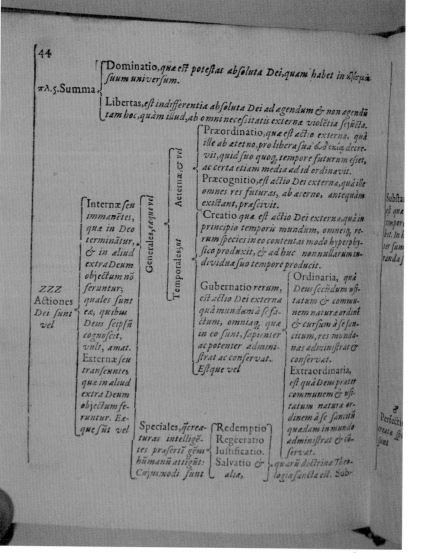

πλ.5.Summa

Dominatio, *quæ est potestas absoluta Dei, quam habet in vsiopa suum vniuersum.*

Libertas, *est indifferentia absoluta Dei ad agendum & non agendu tam hoc, quàm illud, ab omni necessitatis externæ violetia seiuncta.*

ZZZ
Actiones
Dei sunt
vel

Internæ seu immanetes, quæ in Deo terminatur, & in aliud extra Deum objectum nō feruntur; quales sunt eæ, quibus Deus seipsū cognoscit, vult, amat.
Externæ seu transeuntes quæ in aliud extra Deum objectum feruntur. Eæque sut vel

Generales, eæque vel

Aeternæ, & vel

Præordinatio, *quæ est actio externa, quâ ille ab æterno, pro libera sua & d'oxiæ decreuit, quid suo quoq; tempore futurum esset, ac certa etiam media ad id ordinauit.*

Præcognitio, *est actio Dei externa, quâ ille omnes res futuras, ab æterno, antequam existant, præsciuit.*

Temporales, ut

Creatio *quæ est actio Dei externa, quâ in principio temporis mundum, omnesq; rerum species in eo contentas modo hyperphysico produxit, & adhuc nonnullarum indiuidua suo tempore producit.*

Gubernatio *rerum, est actio Dei externa quâ mundum à se factum, omniaq; quæ in eo sunt, sapienter ac potenter administrat ac conseruat.*
Estque vel

Ordinaria, *quâ Deus secūdum vsitatum & communem naturæ ordinē & cursum à se sancitum, res mundanas administrat & conseruat.*

Extraordinaria, *est quâ Deus præter communem & vsitatum naturæ ordinem à se sancitū quædam in mundo administrat & cōseruat.*

Speciales, ā creaturas intelligētes præsertī genus hūmanū attigūt: Cujusmodi sunt

Redemptio
Regeeratio
Iustificatio.
Salvatio & alia,

quaru doctrina Theologiæ sanctæ est. Sub-

Substa
est qua
tempore
bit. In h
ter sum
randa

Perfecti
creatæ sp
sunt

Plate 9: *Ogdoas Scholastica*, p. 44.

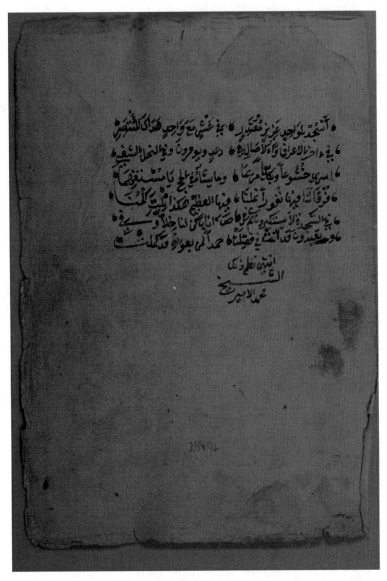

Plate 10: The first page of the *Mahdiyya Qu'ran*: this is a 346 page document of some beauty, first closely studied by Brockett: "fol. 346 (ff. 247, 341, 342 cancels); 234–238 × 160–164 mm.; written area 170–175 × 100–102 mm.; 13 lines per page; laid paper; east Sudan naskh hand in black ink, finely vocalised also in black, with recitative notation, verse-dividers and sura-titles in red; frequent marginal notes again in red; no decorations; strong leather loose-cover binding artistically tooled, ending in an envelope-flap; dated 1299 (1881AD)" [Brockett, 1987].

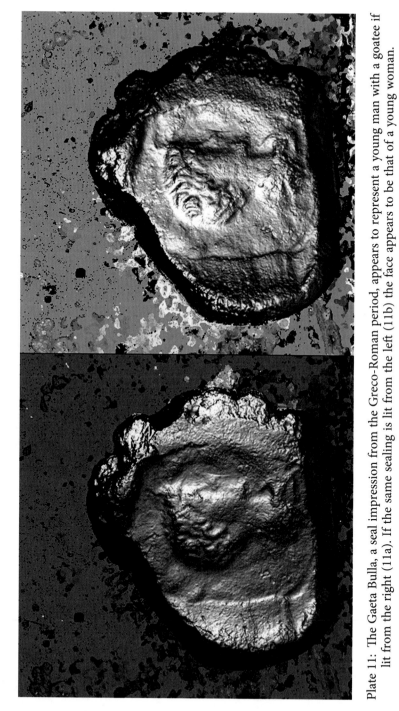

Plate 11: The Gaeta Bulla, a seal impression from the Greco-Roman period, appears to represent a young man with a goatee if lit from the right (11a). If the same sealing is lit from the left (11b) the face appears to be that of a young woman.

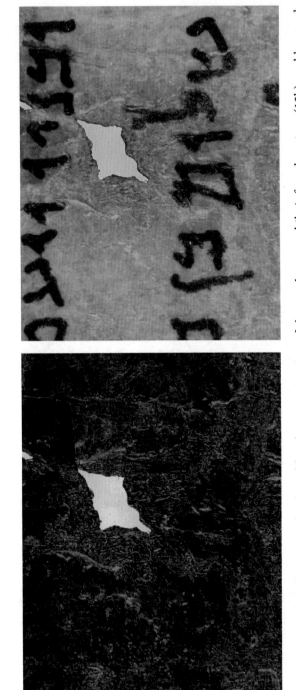

Plate 12: Visible light capture (12a) enables close examination of the parchment, while infrared capture (12b) enables much more clear examination of the written characters.

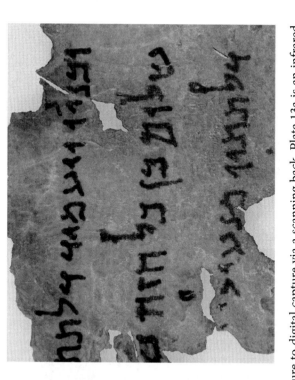

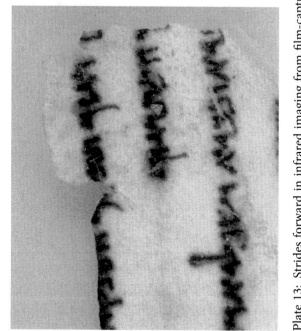

Plate 13: Strides forward in infrared imaging from film-capture to digital capture via a scanning back. Plate 13a is an infrared image of a Dead Sea Scroll, taken in 1988 using large format high speed infrared film. Plate 13b is an infrared digital image of a Dead Sea Scroll at about the same magnification but with significantly higher resolution, taken in 2008, employing a large format camera with scanning back.

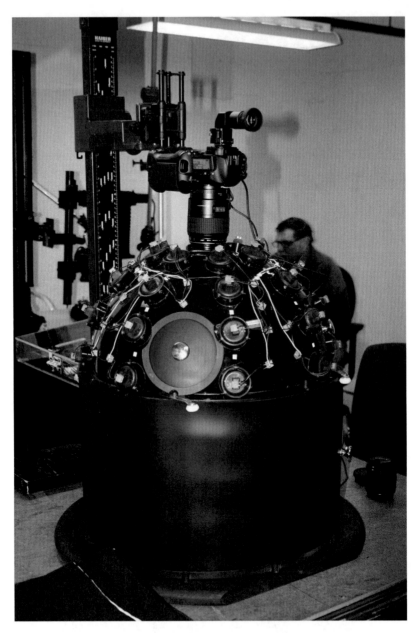

Plate 14: Light dome utilized for capturing images using Reflection Transformation Imaging (RTI) technology. An RTI image is created by taking many digital photographs, each from a different light angle. The images are then combined using special software to build a master image in which one can view an object from any light angle by moving a mouse-driven cursor around the image in real time.

Plate 15: Texture information that can be gained using RTI image technology on a Dead Sea Scroll fragment on animal skin (15a) and a papyrus fragment (15b).

Plate 16: Image capture of a plate of Dead Sea Scroll fragments at the Bibliotheque Nationale de France in Paris using 'the Slider'. In this type of technology, the sensor in a digital scanning back remains stationary while a platform
with the artifact moves past the lens and is captured by the digital software.

Plate 17: In the cylinder seal above, notice how the artisan incised the legs of the mythical beasts (bottom detail), then changed his mind, sanded out the first incisions and recarved the legs in a different, slightly wider stance.

Plate 18: The InscriptiFact spatial search allows researchers to click and drag a box on a reference image and retrieve all images that intersect the box.

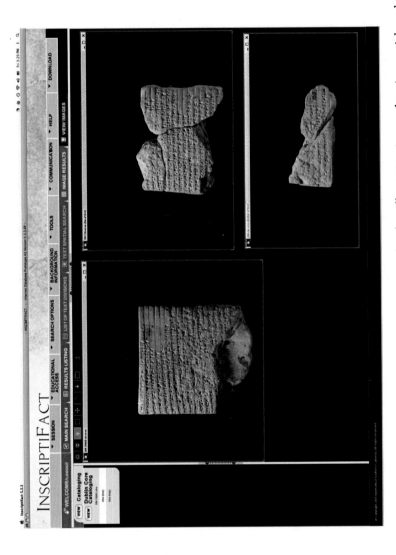

Plate 19: InscriptiFact is designed to incorporate, integrate and index all existing image data in a quick and intuitive fashion regardless of what repository or collection the artifact (or fragments, thereof) exist in. In this example, the original table on which an ancient myth was written was broken, and pieces ended up in two different museums.

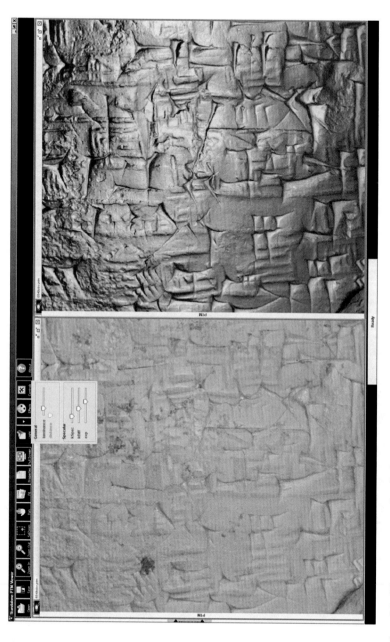

Plate 20: The same image loaded into two side-by-side frames in full-screen mode. In (10a) the image of the artifact is viewed in its natural color, while in (10b) the artifact is viewed in specular enhancement mode, thereby appearing as though it had a highly reflective surface which heightens the texture of the artifact and thus the detail of the text.

Plate 21: Comparing conventional images along-side RTI images in full-screen mode. Illustrated here is an Aramaic tablet from Persepolis, ancient Persia, with a seal impression. The images on the left are visible light and infrared images taken with high-end digital scanning back. The images on the right are versions of an RTI image, one showing the natural color of the object, the other using specular enhancement.

Plate 22: In this RTI image of an Elamite tablet from Persepolis, ancient Persia, two light sources have been selected for viewing the cuneiform text of the tablet. Note that the position of the lights is indicated by a green dot in the upper left of the image, and a red dot on the right side of the image. Using two lights in this way represents the photographic principle of 'key and fill'—one light to highlight shadows and a second light to fill in the shadows so that data will not be obscured.

Plate 23: Codex Sinaiticus, q35, bl4b. Prepared leaves of animal skin. Fourth century. Codex Sinaiticus Project, www.codexsinaiticus.org. AM.59150.l531. q35.bl4b.

Plate 24: Codex Sinaiticus, q37, bl2a.

Plate 25: Codex Sinaiticus, q37, bl3b.

Plate 26: Second Life®: Our Lady Of Peace.

Plate 27: Second Life®: United Methodist Church.

Plate 28: Second Life®: Koinona Meditation Centre.

Plate 29: Second Life®: Hajj.